MAMMAL

MAMMAL

DK

LONDON, NEW YORK,
MUNICH, MELBOURNE, DELHI

Dorling Kindersley Limited

Category Publisher Jonathan Metcalf
Deputy Art Director Bryn Walls
Managing Editor Liz Wheeler
Managing Art Editor Phil Ormerod
Senior Editor Angeles Gavira
Senior Art Editor Ina Stradins
Production Controller Liz Cherry

CONSULTANTS AND CONTRIBUTORS

Dr Paul Bates, David Burnie, Mic Cady, Jonathan Elphick,
Dr David Harrison, Dr Christopher A. Norris, Steve Parker,
Malcolm Pearch, John Woodward

Produced for Dorling Kindersley by

studio cactus ©

13 SOUTHGATE STREET WINCHESTER HAMPSHIRE SO23 9DZ

DESIGN TEAM
Senior Art Editor Sharon Moore
Designers Dawn Terrey, Laura Watson
EDITORIAL TEAM
Senior Editor Kate Hayward
Editors Polly Boyd, Elizabeth Mallard-Shaw

First published in 2003 by
Dorling Kindersley Limited,
80 Strand, London WC2R 0RL

2 4 6 8 10 9 7 5 3 1

A CIP catalogue record for this book is available
from the British Library.

ISBN 0-7513-3933-4

Colour reproduction by Colourscan
Printed and bound by L.E.G.O., Italy

For our complete catalogue visit

www.dk.com

CONTENTS

FOREWORD

At the end of the 18th century, the great taxonomist Carl Linnaeus invented a system for classifying animals, calling warm-blooded animals that feed their young with milk Mammalia. This term covers all the animals that we now think of as mammals, but within this group there is great diversity: the marsupials, or pouched mammals, of Australasia and South America; the egg-laying monotremes (the echidnas and duck-billed platypus) of Australia and New Guinea; and the placental mammals, which make up the rest – including ourselves.

Mammals are found in every environment and climate throughout the habitable world, and their evolution and behaviour is inextricably linked with their natural surroundings. Many species - especially such carnivores as the wolf, the fox, and the brown bear - are highly adaptable. Other species - for example, the grassland antelopes of Africa – are specialists. It is fascinating to think that if our early human ancestors had not been supremely adaptable and had not learnt how to make clothing and light fires, our species would probably still be restricted to the tropical climate in which we first evolved – as is our nearest relative, the chimpanzee.

This book is both an explanation and a celebration of the enormous diversity and incredible beauty of the wild mammals in our family. While we should not forget that without domesticated species such as cows and sheep there would be no agriculture and no civilizations, we should remember that the lives of our wild cousins are closely intertwined with our own, and their survival is largely in our hands.

JULIET CLUTTON-BROCK

SHORT-NOSED ECHIDNA

RED KANGAROO

WHAT IS A MAMMAL?

WARM, FURRY, ACTIVE, ALERT, ADAPTABLE. THESE ARE THE FEATURES
MOST ASSOCIATED WITH MAMMALS. OF ALL THE MAIN ANIMAL
GROUPS, MAMMALS ATTRACT OUR ATTENTION AND STIMULATE
OUR CURIOSITY MOST – BECAUSE WE ARE MAMMALS TOO.

MAIN MAMMAL FEATURES

Mammals are vertebrates that possess
a unique characteristic from which
the group's name is derived: the female
mammal has mammary glands, specialized
body parts on the chest or abdomen that
produce milk. During the early stages
of life, all baby mammals are nourished
on this milk.

In addition to this defining feature,
most mammals share a number of other
characteristics. For example, they are the
only animals that have some covering of
fur or hair. There are exceptions to this: a few, such
as whales, are not at all furry; they are almost hairless.

Mammals are also warmblooded, although the
term is slightly misleading because on a hot
day in the desert a "coldblooded" lizard has a
higher body temperature than that of a gerbil.
A more accurate term to describe mammals is
endothermic, which means that they generate
their own heat within the body by slowly
burning energy-rich food substances; this
process is internally controlled so that
body temperature stays constantly
warm (homeothermy) and does
not depend upon the external conditions.
Coldblooded animals, such as reptiles and fish, have
to absorb heat from outside in order to maintain
optimum body temperature so, if environmental
temperatures fall their body temperatures fall.

ORIGINAL MAMMAL?
The first mammals appeared
over 200 million years ago
and probably resembled
today's tupaids or tree shrews.

If the temperature falls to a much colder
degree than normal, their muscles cease to
work, so the coldblooded animal's activity
level depends on external conditions.

The mammal's ability to control body
temperature and therefore stay active in
almost any environment makes mammals
supremely adaptable. Different species
have evolved to survive in the coldest
polar regions, on the highest mountains,
in the hottest deserts, in deep lakes and
dark caves, and in virtually every other
habitat, including the air and the open
ocean. Even the group of animals that is most numerous
– the insects – cannot rival the mammals' global
domination of the land and occupation of sea and sky.

COPING WITH CHANGE

Mammals are adaptable in two distinct ways. Their
bodily designs and features have evolved – over
thousands and millions of years – to suit different
conditions and lifestyles. They are also adaptable over
shorter time spans – days or even hours – in that they
adjust their behaviour according to the circumstances.

Compared to other animals, a typical mammal
has a larger brain in proportion to body size. Also,
the parts of the brain involved in what we term
intelligence – such as learning by modifying actions
to take account of experience – are better developed than
in other animals. Such adaptability in body, brain, and
behaviour is a major reason for the mammals' success.

AFRICAN ELEPHANT

MAMMAL GROUPS

Monotremes, such as echidnas,
lay eggs from which babies
hatch. Marsupials, such as
kangaroos, give birth to tiny,
poorly developed young that
continue growth in a pouch.
Placentals, which include
elephants, give birth to well-
formed offspring. In all cases,
the young suckles after birth.

EXTREME ADAPTATION

Whales, such as this sperm whale, are the
mammals that have evolved furthest from the
original, small, furry, shrew-like body design.

RIGID AND LIMITED

Unlike most mammals, naked mole-rats of East Africa are very rigid in their behaviour. They live in communal burrows in which each individual carries out one task that contributes to the survival of the whole colony. The system can be compared to that of insects such as ants: their roles are precise and specific, and just one female, the "queen", produces young.

SPECIFIC TASKS
Some "worker" naked mole-rats dig tunnels for roots and other food, while others care for the young and clean out chambers.

FLEXIBLE, EXTENDED BEHAVIOUR
Apes such as the orang-utan show exceptionally intelligent and adaptable behaviour. A youngster spends up to 8 years with its mother, learning the ways of the forest.

HIGH-ENERGY LIFESTYLES

To maintain its constant body temperature a mammal must convert some of its food into an ongoing supply of heat energy. To allow for this it must eat much more food than a coldblooded animal of similar size must consume. For example, the lion and crocodile have a similar diet, chiefly the flesh of large prey such as gazelles and antelopes. But the lion must eat about ten times more than the crocodile does in order to have sufficient energy to fuel body warmth.

Nutritional quality is another factor. A carnivore's diet is much richer in nutrients and energy than the poor-quality, difficult-to-digest grasses, leaves, and similar plant matter eaten by herbivorous mammals. So, for example, zebras must eat more food than lions do in order to extract the energy they need for

LEAF-GRABBER TONGUE

The rare okapi, a West African forest relative of the giraffe, curls its 50cm (20in) tongue around twigs in order to pull the leaves into its mouth.

MEATSCRAPER TONGUE

The lion's front teeth are for stabbing, not nibbling. So it uses its very rough-surfaced tongue as a rasp or file to tenderize tough meat and scrape off bits to swallow.

their ever-alert, ready-to-flee lifestyles. Overall, the mammals' high food intake means that they play hugely important roles in the food chains of nature – both as plant-eaters and meat-eaters.

Some mammals are capable of controlling their body temperature to such an extent that in cold conditions they can conserve energy by hibernating. This involves entering a very deep sleep during which their body temperature falls to just a few degrees Celsius. This does not mean that they become coldblooded: a hibernating mammal is still controlling its body temperature internally; coldblooded animals must rely on external conditions.

SPECIALIST DENTITION

The contrast in lifestyle between mammalian herbivores and carnivores shapes many bodily features, especially the head. A glance at a mammal's mouth and teeth readily identifies its diet. A herbivore such as a zebra, gazelle, or giraffe, has long jaws worked by powerful, high-stamina muscles. Small incisor teeth at the front of the mouth nip off plant matter, aided by the long tongue and prehensile lips, while batteries of broad-topped molar teeth chew for many hours. A carnivore such as a cat, weasel, or otter has a shorter muzzle but a wider gape, giving extra leverage to the bite so that the large-pointed canine teeth can be stabbed forcefully into another animal. The ridged cheek teeth work like shears to slice meat, gristle, and bone.

Two other major types of mammal dentition belong to the rodents and insectivores. About two out of five mammal species are rodents, such as rats, mice, squirrels, beavers, and porcupines. Their specialist feature is a pair of long, chisel-like incisors at the front of both the upper and lower jaws. These teeth are ideally positioned to crack and gnaw tough or shelled plant parts such as nuts, seeds, and roots. Unlike most mammalian teeth, the rodent's incisors continue to grow through life to compensate for the wear inflicted by hard food items. Insectivores, such as shrews, moles, and hedgehogs, have numerous small, sharp teeth for grabbing and chomping not only insects but all kinds of small creatures, including worms, spiders, and slugs.

Omnivorous mammals show a combination of tooth forms, which enable them to eat a wide range of plant and animal food. Among these omnivores are primates such as monkeys and apes.

LIMBS AND LOCOMOTION

Mammals demonstrate a vast range of adaptations in the ways that they move about. Most mammals have four limbs, but they use them in various ways. Some use them almost entirely for walking or running. Among the most highly adapted runners are hoofed mammals. Their long, slim legs have fewer toes than the standard mammalian pattern of five per limb, and each toe is capped with a tough, hardwearing but lightweight horn (hoof). In zebras and horses, the toes of each limb have evolved into one single hoof.

Most of the muscle bulk that moves a mammal's limb is concentrated in the upper region, where it joins the body. This allows the lower section's structure to be comparatively simple and lightweight, which maximizes manoeuvrability or enables it to swing to and fro with maximum speed when running.

Some mammals, such as monkeys and rodents, have multi-purpose limbs with grasping hands and feet for climbing trees and/or handling food. Two very different kinds of mammal – bats and cetaceans – have evolved limbs with broad, wide surfaces. In bats, the forelimbs have become adapted as wings, each with a very thin, stretchy, skin-like membrane, the patagium, held out by elongated finger bones. In cetaceans, such as dolphins, the limbs have become flippers for swimming.

SAFETY IN NUMBERS
Wildebeest (gnu) gather as massed herds for
seasonal migrations to new grazing. Their vast
numbers deter many predators, although the
young and the sick will always be targeted.

SOCIAL SPECTRUM

Throughout the animal kingdom, the vast majority
of creatures live solitary lives. They meet up with
others of their kind only to breed. In contrast, among
the insects, certain kinds of ants, wasps, and termites
must form colonies and cannot survive alone.
Mammals show these two extremes of social
behaviour, and all gradations between. In general,
most small mammals – and also most meat-eaters,
from bears and tigers to tiny shrews – usually live
alone, except during courtship and when a mother
is rearing her offspring.

HERBIVORES

Many kinds of larger herbivores, such as zebras, deer,
giraffes, and the bovids – antelope, gazelles, cattle, sheep,
and goats – live in associations known as herds. Their
individual interactions, other than those
related to breeding, are not
especially complex. But the
grouping provides several
advantages for survival. In
a herd there are hundreds
of eyes, ears, and noses
alert for danger, and when
one individual detects a
threat, its startled reaction
warns all others near by.
Any predator will be
swamped by the numbers

FUR COAT

Fur or hair provides insulation from cold and protection
against physical damage. It also acts as camouflage
for mammals whose survival strategy is
to blend into their surroundings.

INNER AND OUTER
Like many mammals, the mink has a dense
insulating undercoat of soft hairs and a protective
outer (guard) coat of long, coarse hairs.

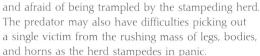

FEMALES IN CHARGE
The spotted hyena group
is female-dominated. Members
live in a communal den, hunt
together, and jointly defend
their territory and youngsters.

and afraid of being trampled by the stampeding herd.
The predator may also have difficulties picking out
a single victim from the rushing mass of legs, bodies,
and horns as the herd stampedes in panic.

Body adornments on these large herbivores –
especially those on the head, such as horns, tusks,
and antlers (see opposite) – are used both as practical
tools for feeding and defence and as visible symbols
of sexual maturity and readiness to breed. This is
especially important among males, firstly as a means
of intimidating rivals while seeking access to females
and then to impress the females as a prelude to mating.

CARNIVORES AND PRIMATES
Some of the most close-knit and complex social
groupings are among the carnivores and primates.
Certain kinds of foxes form long-term pairings. Other
members of the dog family, such as wolves and bush

dogs, are well known as pack-dwellers. Wolves have a detailed hierarchy system, where only the dominant female and male, the "alpha pair", breed. Subordinate pack members bring food for the offspring, while adult members will work together as a team to bring down prey larger than themselves, such as an elk. But wolf behaviour is very flexible and when smaller prey such as hares and rabbits is plentiful pack members split up to hunt on their own. Hyenas, too, live and hunt in groups, whereas the aardwolf, and most cats, are solitary.

Among the major mammal groups, it is in the primates – monkeys, apes, lemurs, and bushbabies – that sociality is most widely developed and most varied. It ranges from the pair-for-life system that is seen in most gibbons, to the extended family unit system followed by gorillas, to the much larger and looser aggregations typical of baboons and capuchins. Chimpanzees, in particular, form smaller "friendship" groups within subtroops. These often coalesce into larger troops, which in turn may come together to form even bigger bands. In all

these groupings, members will gather together to distract or frighten an enemy, and will then spread out again to feed once the threat has passed. If one member of the troop finds a plentiful supply of food, the others come around to share. In some cases, adults without offspring, known as helpers, assist parents by "babysitting".

SMALL AND SECRETIVE
In herds, the rabbit-sized, spotted chevrotains would be noticeable and vulnerable because of their small size. So these mouse deer live solitary, secretive lives in Asian rainforests.

HEAD ORNAMENTS

Mammal head adornments are made of various materials. Rhino "horns" are compacted, matted hair; elephant tusks are enlarged teeth; deer antlers are bone-based and shed yearly; bovid horns are true horn and grow throughout life.

**NOSE HORNS:
RHINOCEROS**

**IVORY TUSKS:
ELEPHANT**

**BONY ANTLERS:
CARABOU**

**HORNS: SCIMITAR–
HORNED ORYX**

NOTABLE FACE SHAPES

The relative importance of a mammal's senses, combined with the adaptations of its jaws and teeth to manage particular food types, combine to fashion the overall shape and proportions of the face.

GIANT ANTEATER

CHIMPANZEE

COYOTE

HARBOUR SEAL

DAMA GAZELLE

SHAPED FOR A PURPOSE

During evolution, the forces of natural selection have shaped and fashioned almost every part of a mammal's body to maximize chances of survival and breeding in its habitat. The huge range and diversity of habitats around the globe have resulted in a similarly wide variation in shape and size of body features.

In particular, the mammal's sensory parts show great variety. Nocturnal mammals such as mice, rats, and cats have big eyes so as to be able to pick up even the faintest amounts of light. Seals live in the reasonably clear ocean, but have large eyes so they can spot prey in water, and danger, such as a polar bear, on land. In contrast, visibility is so poor in the muddy murk of slow-moving tropical waterways that river dolphins have tiny, almost non-functioning eyes. They use another sensory modality to find their way and their prey – sound. The dolphin makes high-pitched clicks and squeaks and detects any returning echoes, which it analyses to establish the direction, distance, and size of objects near by.

BUILT–IN EXTRAS

Bats use a similar, sound-based system, called echolocation, to navigate and feed as they fly even on the darkest nights. In some bats, including the various long-eared species, the ears are far larger than the head. Indeed, many mammals have large external

COOL EARS
The fennec (p.42) is the smallest fox but it has huge ears, both to hear beetles and other prey scrabbling in the desert sand, and to release excessive heat.

ears shaped like funnels or dishes, which they can swivel so as to catch the slightest sound and pinpoint its location, all without moving the head.

In some cases acute hearing is not the only function of huge ears. Mammals of hot habitats like deserts and tropical grasslands avoid overheating by using their ears as "radiators" to get rid of excess body warmth. These include tiny gerbils and jerboas, desert hares and jackrabbits, fennec and kit foxes, and the largest of terrestrial mammals, the elephants. On their open plains, African savanna elephants endure much more hot sun than Asian elephants do in their shady forest habitats – which may be one reason why the former has relatively larger ears.

A SENSE OF SMELL
The facial shapes of many mammals are dictated by the design and relative size of the sensory parts, particularly the eyes, ears, and noses. A long snout or muzzle, which houses a large chamber within which invisible airborne particles of odours can be trapped and detected, indicates a strong sense of smell. By contrast,

MURKY WATERS
The baiji or Chinese river dolphin navigates by touch and sound in the cloudy water of its murky home.

EARS BIG AND SMALL

Different species of lagomorph mammals – mainly rabbits and hares – show a range of ear sizes. These are directly related to habitat and lifestyle. The species that live above ground in the hottest, driest terrain have the largest ears. Species from colder places and/or those that dwell in burrows have shorter ears.

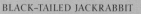

BLACK-TAILED JACKRABBIT
This hare's outsized ears release heat during the scorching summers of North America's southwestern deserts.

predominantly visual species, including monkeys and apes, have flattish faces and their olfactory sense is less acute. Mammals that rely greatly on smell include not only larger herbivores such as deer and antelopes, but also many smaller, nocturnal or underground types – notably mice, shrews, moles, and mole-rats. Frequently, the quivering muzzle fulfils two sensory tasks: while it sniffs for smells, its bristling whiskers are also ultra-sensitive to touch. Like smell, touch is a sense that works in darkness.

ADAPTATION TO EXTREMES

The most severe habitats require the most exceptional changes to the basic mammalian body plan. In the coldest regions of the world there have been several evolutionary responses to the problem of heat loss from a mammal's warm body. On top of the skin, fur coats are much longer and denser than elsewhere (with more hairs per unit area). Under the skin, a layer of a fatty substance, blubber, helps to keep in warmth, especially in seals, whales, and other aquatic mammals. The body's extremities, which lose heat fastest, are much decreased in size. For terrestrial mammals, these include the ears, muzzle, and tail. Comparisons of related species from cold and hot habitats, such as the Arctic hare and black-tailed jackrabbit (see opposite and p.71) or the Arctic fox (p.90) and fennec fox (see opposite), show how environmental extremes help to shape their mammal inhabitants.

A LEAP IN THE DARK
The woodmouse is well equipped for moving at night, with its quivering nose, long twitching whiskers, large ears, and huge beady eyes to catch the dimmest ray of light.

CHANGING ITS SPOTS
The black panther is a leopard found mainly in dense Asian rainforest, where it is better camouflaged than the spotted form.

THE LONG-EARED BAT
This bat's ears are four times as long as its head, receiving the bat's own ultrasound echoes and the flutterings of prey such as moths.

USEFUL RESOURCES

The behavioural adaptability of mammals such as monkeys and apes allows them to vary their means of survival. In particular, capuchin and macaque monkeys exploit almost any type of food source and readily take advantage of any local resources that might provide shelter.

HOT SPA
In winter "snow monkeys" (Japanese macaques) warm up in hot springs in the far limits of their region.

In extreme conditions, bodily adaptations are sometime supplemented by behavioural ones. In exceptional heat, desert species from kangaroos to kangaroo-rats search out cool shade, perhaps under a tree or a rocky overhang, or in a cave. In the cold winters of the northern lands, mammals learn to visit local hot springs to revive themselves with a warming drink or, in the case of the Japanese macaques (see above), a hot bath.

WHITE AND WARM
The newborn harp seal's thick coat is white for camouflage on the ice. In conjunction with the blubber beneath the skin, it keeps the seal warm in extreme, sub-zero temperatures.

EVOLUTION

MAMMALS HAVE A LONG AND COMPLEX PREHISTORY. THEY
APPEARED ON EARTH JUST AFTER THE FIRST DINOSAURS
AND THEN, AFTER THOSE GREAT REPTILES BECAME EXTINCT,
SPREAD AND DIVERSIFIED AT TREMENDOUS RATES.

EARLY AND TINY
One of the earliest mammals was *Morganucodon*.
It was just 10cm (4in) long, and its fossils come from
western Europe and eastern Asia.

THE THIRD MAMMALIAN FEATURE

Two of the defining features of a mammal are a furry
or hairy body and milk or mammary glands in the
female, which produce milk to feed her offspring.
The evidence used by experts investigating origins
and evolution is mainly fossils – the preserved
remains of bones, teeth, and other hard body
parts. However, fur or hair fossilizes very
rarely, and mammary glands not at all.
Luckily, mammals have a third defining
feature, identifiable in both living kinds and in
fossilized remains. This is the presence of three
tiny bones, or ossicles, (called the hammer,
anvil, and stirrup) deep inside the ear. Over
millions of years of evolution, two bones in
the lower jaw of the synapsids – reptile-like
mammal ancestors – moved to the ear to become
part of the sound-amplifying system that is found
only in mammals. The incorporation of these bones
into the ear meant that the lower jaw, which had
previously been three or more bones in reptiles,
became just one bone (the mandible). Alongside

LATER AND BIGGER
The imperial mammoth
of North America weighed
a million times more than
Morganucodon (see above).
It lived from about 2 million
to 10,000 years ago.

associated changes in the skull, teeth, and other skeletal
parts, these features allow us to identify mammals in
the fossil record.

SMALL BEGINNINGS

The earliest mammals date back over 200 million
years, to near the end of the Triassic Period.
At this time dinosaurs were relatively new
and were changing and spreading fast.
Mammalian evolution was slower and
less adventurous, with most early kinds
being small, nocturnal insect-eaters –
similar in appearance to today's tree shrews
(p.10) and in habits to shrews. While dinosaurs
ruled the land over the next 140 million years,
several mammal groups came and went –
hardly any of them larger than a pet cat.
The most ancient of the three living groups is the
monotremes, which appeared in the Early Jurassic
Period (see diagram below), when dinosaurs began
to attain immense size. The marsupials evolved around
the middle of the next period, the Cretaceous, closely
followed by today's largest group, the placentals.

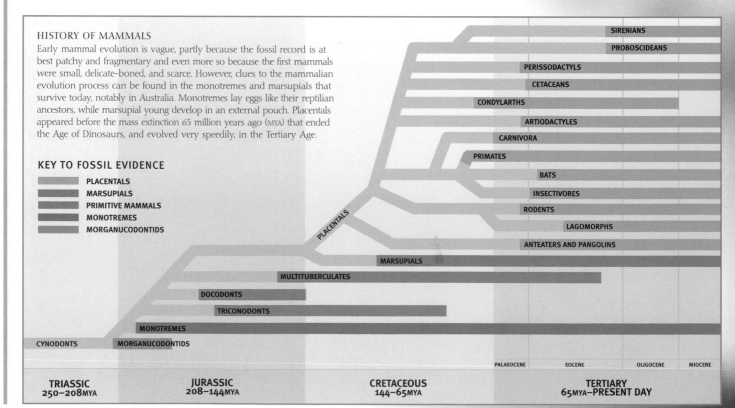

HISTORY OF MAMMALS
Early mammal evolution is vague, partly because the fossil record is at
best patchy and fragmentary and even more so because the first mammals
were small, delicate-boned, and scarce. However, clues to the mammalian
evolution process can be found in the monotremes and marsupials that
survive today, notably in Australia. Monotremes lay eggs like their reptilian
ancestors, while marsupial young develop in an external pouch. Placentals
appeared before the mass extinction 65 million years ago (MYA) that ended
the Age of Dinosaurs, and evolved very speedily, in the Tertiary Age.

KEY TO FOSSIL EVIDENCE
- PLACENTALS
- MARSUPIALS
- PRIMITIVE MAMMALS
- MONOTREMES
- MORGANUCODONTIDS

SIRENIANS
PROBOSCIDEANS
PERISSODACTYLS
CETACEANS
CONDYLARTHS
ARTIODACTYLES
CARNIVORA
PRIMATES
BATS
INSECTIVORES
RODENTS
LAGOMORPHS
ANTEATERS AND PANGOLINS
PLACENTALS
MARSUPIALS
MULTITUBERCULATES
DOCODONTS
TRICONODONTS
MONOTREMES
CYNODONTS
MORGANUCODONTIDS

PALAEOCENE | EOCENE | OLIGOCENE | MIOCENE

TRIASSIC
250–208MYA

JURASSIC
208–144MYA

CRETACEOUS
144–65MYA

TERTIARY
65MYA–PRESENT DAY

WOLF OF TODAY
The grey wolf is the latest in a long line of wolf-type predators. They did not all belong to the carnivore group: *Hyaenodon*, from 30 million years ago, for example, belonged to the now extinct Creodonta.

CYNODESMUS

CANIS DIRUS

WOLF OF YESTERDAY
Cynodesmus was one of the first of the Carnivora to resemble modern dogs and wolves. Dire wolves, *Canis dirus*, are known from hundreds trapped in tar pits over 20,000 years ago in what is now Los Angeles, USA.

MAMMALS ON THE MOVE

The world map was once very different from today's. The supercontinent Gondwana (now South America, Australia, Africa, India, and Antarctica), which formed about 650 million years ago, was in the process of breaking up when the first mammals appeared. Each land mass took its own selection of mammals, dictating which groups live where today. Marsupials, for example, survived well in Australia, isolated from the major placental groups.

North Asia
South China
Gondwana
North America
North Europe
Siberia
Baltica

ALL AS ONE
The same animal found on separate continents today is an indication that prior to continental drift these landmasses were joined.

CHANGING FORTUNES

More than half of the orders (major groups) of mammals that have ever existed are now extinct. They flourished for a time, then succumbed to changes in climate, or competition from newer mammals for food or living space. These varied fortunes were especially marked among the hoofed mammals, or ungulates. For example, South America was isolated from Africa and North America by ocean for millions of years, and lacked large placental carnivores. From about 60 million to 5 million years ago many hoofed mammal groups, including litopterns and notoungulates, evolved hundreds of species there. These resembled not only today's ungulates, such as horses, rhinos, tapirs, camels, and antelopes, but also rabbits, sloths, and bears. Yet not one species of these earlier South American mammals, known collectively as meridi-ungulates, survived the competition from North American species.

EARLY SUCCESS

The fossil record shows that numerous mammal orders appeared very rapidly after the dinosaurs' demise in the Early Tertiary Period from

EARLIEST BAT
Icaronycteris from North America was an early bat. Similar types lived in Europe, Africa, and Australia.

LITTLE BROWN BAT
There are now nearly 1,000 species of bats, making them the second most species-rich mammal group.

MAMMALS TAKE TO THE AIR

Bats such as *Icaronycteris* appear abruptly in the fossil record some 55 million years ago as well-evolved, agile fliers that had spread to several continents. Little or no earlier fossil evidence has survived. Modern species have fewer teeth, and have lost the claw on the second finger.

FOSSIL BAT
Bat bones are so thin they rarely fossilize. This 50 million-year-old *Palaeochiropteryx* is from Messel, Germany.

approximately 65 to 55 million years ago. During this burst of evolution, many groups were relatively short-lived, lasting just a few million years. But a few became well established and continue successfully to the present day. They include two of the most specialized or highly evolved of all mammal groups - bats in the air, and whales and other cetaceans (p.203) in the ocean.

DEMISE AND DOWNFALL

Other mammals enjoyed much greater success in previous times than today. Among the odd-toed ungulates (perissodactyls), the rhinoceroses were once tremendously varied and widespread, with hundreds of species. Some resembled horses, while others looked like hippopotamuses or pigs. Although today's rhinos all have nose horns, most of the prehistoric ones did not.

The hornless *Paraceratherium* (see right) was a type of rhino, living in Asia. With a head-body length of 9m (34ft), shoulder height of over 5m (16ft), and weight of 15–20 tonnes (33,000–44,000lb), it was the largest land mammal ever. From about 5 million years ago the rhino group began to fade. Early humans probably had a hand in its most recent loss – the woolly rhino, *Coelodonta* (see right), of the last ice age. It survived the great ice ages of the last million years in Eurasia – it was depicted in European cave paintings nearly 30,000 years ago, and frozen corpses from Siberia are less than 10,000 years old. Today the rhino group has only five species, and their demise continues at the hands of humans. All rhino species are highly threatened, despite being protected, particularly the northern white rhinoceros, *Ceratotherium simum cottoni*.

While new living species do continue to evolve, the rate at which this is occurring falls far behind current rates of extinction and, unless humans radically alter their ways, the catastrophic rise in the number of species becoming extinct can only continue.

PARACERATHERIUM

COELODONTA

RHINOS OF THE PAST
The hornless rhino *Paraceratherium* lived in Asia about 25 million years ago. The woolly rhino, *Coelodonta*, survived the extreme changes in the Earth's climate, but not hunting by humans greedy for its horn.

SURVIVING – BUT ONLY JUST
Limited to Africa and Asia, today's rhinos look very different from many of their prehistoric cousins. The hook-lipped (black) species is the second-rarest, numbering about 3,100.

THE ROLE OF MANKIND

HUMANS ENGAGE IN ALMOST EVERY TYPE OF RELATIONSHIP WITH OTHER
MAMMALS. WE BEFRIEND THEM, TRAIN THEM, WORSHIP THEM, EXPLOIT
THEIR STRENGTH AND MOBILITY, UTILIZE THEIR BODY PRODUCTS – AND,
OF COURSE, EAT THEM.

FROM INTEGRATION TO DOMINATION

In terms of evolutionary time, in the blink of an eye humankind has gone from being an integral part of nature, to dominating it, eradicating huge areas of it, and threatening much of its survival – especially its mammals. The varied and complex interactions between people and mammals have passed through several phases, although they happened at different times in different regions of the world.

The emerging technologies of early people were based around tool use of hammer-stones, rock-scrapers, and, later, hand-axes and spears. The fossils of large mammal bones at prehistoric sites, some more than one million years old, show signs of being cut and smashed by these tools. From 50,000 years ago, progress in weapon design began to accelerate significantly – as did humankind's domination of the environment. Many of these advances occurred in Europe, where settlements were littered with the tusks of elephants and mammoths, the antlers of deer, and the horns of antelopes and gazelles.

THE FIRST OVERKILL?

Early forms of art also began to appear, including Aboriginal rock art in Australia and cave paintings, carvings, and sculptures in Europe. Many depict large mammals, including giraffes, antelopes, hippos, rhinos, horses, elephants, lions, and wolves. Some examples show hunting scenes where this "big game" is pursued

DRAUGHT WORK
Heavy horses were bred to pull ploughs and carts. Since tractors took over most of their work, traditional breeds – here Suffolk Punch – have dwindled in number.

TRANSPORT
As our "best friends", dogs find roles from pulling sleds in snowy lands to sniffing out suspicious substances.

by people or pierced by spears. Even at these early stages of civilization, it seems that larger mammals, in particular, were the subject of both celebration and slaughter.

The fade of the last great ice age 20,000 –10,000 years ago seems to coincide with the disappearance of several large mammal species from northern continents. Woolly mammoths, woolly rhinos, and cave bears are well-known examples. The climate was changing rapidly at the time. However, many experts argue that the spread of humans, with their intelligent hunting strategies and improving weaponry, was the central cause of this, the first wave of our many "mammal overkills".

SETTLING DOWN

From about 10,000 years ago humans entered a major new phase of civilization, as temporary settlements became towns and then cities. The start of agriculture saw crops sown and mammals domesticated. Early forms of selective breeding changed formerly wild and wide-ranging ungulates into less aggressive, more productive livestock. Domesticated sheep (*Ovis aries*), goats (*Capra hircus*), pigs (genus *Sus*), and cattle (genus *Bos*) became part of daily life. These mammals were kept for meat, milk, and hides, and some for wool. The need to venture into the wild on hunting trips was much reduced. Within a few thousand years large populations of humans were living not so much as an integral part of nature – hunting, and gathering – as apart from it, in village squares and town streets.

IMAGERY
Cave and rock art from prehistoric times, such as these paintings from Namibia in southwest Africa, show how people were fascinated by large mammals, partly as sources of food.

IUCN, RED LISTS, AND CITES

IUCN (International Union for the Conservation of Nature and Natural Resources) was founded by the United Nations in 1948 to carry out a range of activities that were aimed at safeguarding the natural world. Part of its work is the regular compilation of the "Red Lists" of threatened species, including mammals. The lists draw together information from more than 10,000 scientists worldwide, and is a global directory of species under threat. In 1973 81 nations signed the treaty CITES (Convention on International Trade in Endangered Species). This aims to control trade in threatened species, either alive or dead, including body parts and products. CITES now has over 120 signatory countries.

NEW FRIENDS

Livestock in the fields still needed guarding and rounding up in order to count, milk, shear, or slaughter them. Wolves (*Canis lupus*) and other wild canids had probably been hanging around campfires for millennia before, 10,000 years ago or earlier, they entered into a more formal partnership when selective breeding and training produced early types of domesticated dogs. These dogs probably played varied roles, as hunting companions, livestock defenders, personal bodyguards – and companion pets. Today the dog remains our number one companion mammal. All pet dog breeds belong to a single species, *Canis familiaris*. Varying from tiny chihuahuas to enormous Saint Bernards and Great Danes, this species shows more variation in breeds and sizes than any other domesticated mammal.

The cat is second only to the dog as our main mammalian pet, but it has a shorter domestic history. It probably originated from North African representatives of the wildcat species, *Felis sylvestris*. From about 5,000 years ago it was tolerated and then encouraged as a killer of rats, mice, and other rodents around grain stores, farmyards, and dwellings. Many pet-lovers today insist that humans have not accepted and trained the cat – others believe the reverse is true.

STRENGTH AND MOBILITY

A horse can run faster than a human, and a water buffalo can pull a heavier load. Through the ages, a variety of large mammals have been domesticated, both as means of transport and as beasts of burden. Horses were probably first ridden in Central Asia, perhaps more than 4,000 years ago. They have played a vast role in shaping the national boundaries of the world by carrying troops and pulling chariot-riders into battle and, in recent centuries, towing cannons and other pieces of war equipment into battle zones. Przewalski's wild horse of Asia (see page 157) may be similar to the original stock from which horses were domesticated. All horses today belong to a single species, *Equus caballus*.

WOOL AND MEAT
Sheep are probably the second most numerous domesticated large mammal after cattle. There are more than 150 million in Australia alone.

LIVESTOCK CARE
Sheepdogs are trained to channel the natural instincts of their wild ancestors to our own wishes and wants – one behavioural aspect of the process of domestication.

Other larger domesticated mammals include water buffalo (*Bubalus bubalis/arnee*), donkeys (*Equus asinus*), and several species of the camel family, including the one-humped dromedary (*Camelus dromedarius*) in North Africa and the Middle East and the two-humped bactrian (*Camelus bactrianus*) in Central Asia. In South America two further members of the camel family were domesticated from about 7,000 years ago, the llama from the wild guanaco (*Lama guanicoe*) and the alpaca, probably from the vicuna (*Vicugna vicugna*).

FERAL MAMMALS

Almost no wild dromedaries are known from Africa and Asia. But this species has been exported to many drier parts of the world, as a pack animal that is suited to desert habitats. In Australia dromedaries have wandered from human influence and now live wild in the outback as feral populations. Horses, dogs, hogs, and several other mammal species have followed the feral route. However, other mammals have much more one-sided relationships with humans. Some, such as brown rats, house mice, red foxes, and raccoons, exploit our own environments and become pests. Others suffer as we destroy their habitats and wild places and drive them towards extinction.

PEST STATUS
Raccoons, foxes, and rats are familiar annoyances in many areas, creating noise and mess, spoiling food stores and perhaps spreading diseases.

ON THE BRINK

Conservative estimates show that one-quarter of all mammal species face some kind of threat. Primates fare worse than most, with more than 40 species or subspecies facing imminent extinction - in some areas whole forests disapper in a week. Some that catch the public imagination, especially mountain gorillas and the orang-utan, can be used to extend awareness to less "glamorous" species, such as bats and marsupial mice.

ROLES REVERSED
Orangs in Southeast Asia are highly threatened: in some areas people now "train" captive-reared individuals to survive in the wild.

MAMMAL CLASSIFICATION

WITHIN THE ANIMAL KINGDOM, ALL MAMMALS ARE INCLUDED IN THE CLASS MAMMALIA, WITH A TOTAL SPECIES NUMBER OF ABOUT 5,000. OTHER CLASSES OF EQUIVALENT RANK INCLUDE BIRDS (MORE NUMEROUS, AT ABOUT 9,000 SPECIES) AND REPTILES (ALMOST 8,000 SPECIES). BY FAR THE LARGEST SUBGROUP WITHIN MAMMALIA IS THE RODENTS, WHICH HAS ALMOST TWO OUT OF FIVE OF ALL MAMMAL SPECIES.

MAMMAL GROUPS

Major groupings within the class Mammalia are known as orders. Biologists differ on the number of orders in the class. Some classification schemes show about 22 orders, others 30, with seven orders for marsupials, three for insectivores, and two for anteaters. The seals and sea lions are also included within the order carnivora by many biologists. The number of species constantly changes, with new discoveries or extinctions.

EGG-LAYING MAMMALS
ORDER Monotremata   FAMILIES 2   SPECIES 5

SHORT-NOSED ECHIDNA

MARSUPIALS
ORDER Marsupialia   FAMILIES 22   SPECIES 292

INSECTIVORES
ORDER Insectivora   FAMILIES 6   SPECIES 365

BROWN LONG-EARED BAT

BATS
ORDER Chiroptera   FAMILIES 18   SPECIES 1,100

ELEPHANT-SHREWS
ORDER Macroscelidea   FAMILIES 1   SPECIES 15

FLYING LEMURS
ORDER Dermoptera   FAMILIES 1   SPECIES 2

INDIAN TREE SHREW

TREE SHREWS
ORDER Scandentia   FAMILIES 1   SPECIES 19

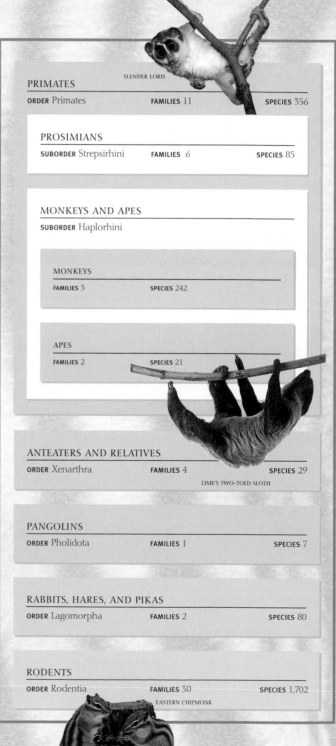

SLENDER LORIS

PRIMATES
ORDER Primates   FAMILIES 11   SPECIES 356

PROSIMIANS
SUBORDER Strepsirhini   FAMILIES 6   SPECIES 85

MONKEYS AND APES
SUBORDER Haplorhini

MONKEYS
FAMILIES 3   SPECIES 242

APES
FAMILIES 2   SPECIES 21

ANTEATERS AND RELATIVES
ORDER Xenarthra   FAMILIES 4   SPECIES 29

LIME'S TWO-TOED SLOTH

PANGOLINS
ORDER Pholidota   FAMILIES 1   SPECIES 7

RABBITS, HARES, AND PIKAS
ORDER Lagomorpha   FAMILIES 2   SPECIES 80

RODENTS
ORDER Rodentia   FAMILIES 30   SPECIES 1,702

EASTERN CHIPMONK

CETACEANS
ORDER Cetacea **FAMILIES** 13 **SPECIES** 83

BALEEN WHALES
SUBORDER Mysticeti **FAMILIES** 4 **SPECIES** 12

TOOTHED WHALES
SUBORDER Odontoceti **FAMILIES** 9 **SPECIES** 71

CARNIVORES
ORDER Carnivora **FAMILIES** 7 **SPECIES** 249

DOGS AND RELATIVES
FAMILY Canidae **SPECIES** 36

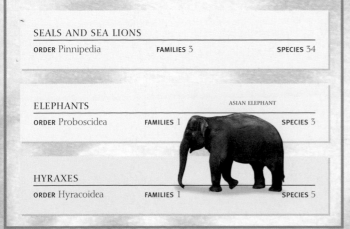
GIANT PANDA

BEARS
FAMILY Ursidae **SPECIES** 8

RACCOONS AND RELATIVES
FAMILY Procyonidae **SPECIES** 20

MUSTELIDS
FAMILY Mustelidae **SPECIES** 67

CIVETS AND RELATIVES
FAMILY Viverridae **SPECIES** 76

HYENAS AND AARDWOLF
FAMILY Hyaenidae **SPECIES** 4

STRIPED HYENA

CATS
FAMILY Felidae **SPECIES** 38

SEALS AND SEA LIONS
ORDER Pinnipedia **FAMILIES** 3 **SPECIES** 34

ELEPHANTS
ASIAN ELEPHANT
ORDER Proboscidea **FAMILIES** 1 **SPECIES** 3

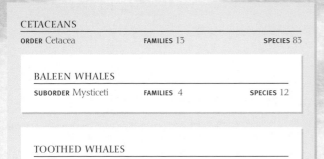

HYRAXES
ORDER Hyracoidea **FAMILIES** 1 **SPECIES** 5

AARDVARK
ORDER Tubulidentata **FAMILIES** 1 **SPECIES** 1

DUGONG AND MANATEES
ORDER Sirenia **FAMILIES** 2 **SPECIES** 4

ODD-TOED HOOFED MAMMALS
ORDER Perissodactyla **FAMILIES** 3 **SPECIES** 19

HORSES AND RELATIVES
FAMILY Equidae **SPECIES** 10

AFRICAN WILD ASS

RHINOCEROSES
FAMILY Rhinocerotidae **SPECIES** 5

TAPIRS
FAMILY Tapiridae **SPECIES** 4

MALAYAN TAPIR

EVEN-TOED HOOFED MAMMALS
ORDER Artiodactyla **FAMILIES** 10 **SPECIES** 225

PIGS AND PECCARIES
FAMILIES Suidae and Tayassuidae **SPECIES** 17

BUSH PIG

HIPPOPOTAMUSES
FAMILY Hippopotamidae **SPECIES** 2

CAMELS AND RELATIVES
FAMILY Camelidae **SPECIES** 7

DEER, MUSK DEER, AND CHEVROTAINS
FAMILIES Cervidae, Moschidae, and Tragulidae **SPECIES** 56

PRONGHORN
FAMILY Antilocapridae **SPECIES** 1

GIRAFFE AND OKAPI
FAMILY Giraffidae **SPECIES** 2

CATTLE AND RELATIVES
FAMILY Bovidae **SPECIES** 140

SITATUNGA

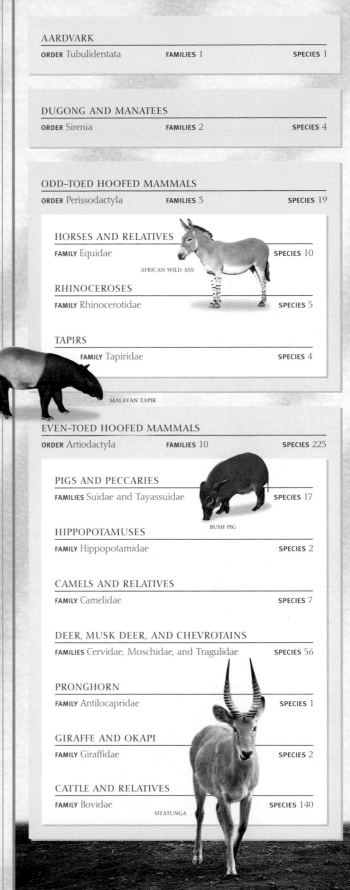

AFRICA

500 MILLION YEARS AGO Africa and Arabia are just south of the equator. Compared with their present-day orientation they are almost "upside down", with what is now southern Africa in a northerly position, at the equator. (Of all the major land masses, Africa has probably drifted around the world least.)

210 MILLION YEARS AGO The supercontinent of Pangaea comprises all major land masses, with Africa (oriented much as it is today) at its centre. The east side of South America fits neatly into Africa's western coastline. Fossils from Lesotho in the south indicate that some of the earliest mammals date from this time (the late Triassic period).

100 MILLION YEARS AGO Africa begins its separation from long-time partner South America, and the South Atlantic Ocean widens. Placental mammals are divided, and diversity increases as they begin to develop in different ways.

6 MILLION YEARS AGO Some ape-like creatures from Africa are evolving into hominids. They probably include our own ancestors.

2–1 MILLION YEARS AGO On the east side of the continent, Earth movements form the Great Rift Valley – a series of huge gashes that extend from the great lakes to the Red Sea.

155,000 YEARS AGO Modern humans appear in East Africa and spread around the world.

PREVIOUS PAGE:
CHEETAH
These are the eyes of the world's fastest animal. A cheetah can sprint at an astonishing 100kph (62mph).

HABITATS OF
AFRICA

WITH THE EQUATOR RUNNING ALMOST THROUGH ITS CENTRE, AFRICA IS MOSTLY WARM. HOWEVER, ITS MOISTURE LEVELS VARY HUGELY, AND AS A RESULT ITS HABITATS RANGE FROM PARCHED TO SATURATED.

A UNIQUE ENVIRONMENT

The world's second-largest continent extends through such a wide range of latitudes – and therefore climates – that its habitats vary considerably from one zone to another. These different habitats support a great diversity of mammals, and some habitats cover such wide areas that they are able to provide sufficient space and food to support very large mammals, such as the African elephant (the world's largest land animal) and the giraffe (the world's tallest). Africa is also home to all the great apes, except the orang-utan, and many other species that are endemic and – in some cases – very rare.

DRY TO WET TO MODERATE

Most of Africa's habitats can be divided into band-like zones that extend west to east across the continent. In the narrow strip that borders the Mediterranean in the far north, the climate is typically Mediterranean: summers are long, hot, and dry; winters are short, cool to moderate, and damp. Immediately south of this zone, the climate suddenly becomes very hot and dry; here, the habitat is characterized by drought and the seemingly endless sand and rocks of the world's largest desert, the Sahara.

Continuing south, gradually increasing rainfall turns the desert to dry scrub, then to shrubby grassland, and on to scattered woodland. In the centre, the combination of rain and heat creates the vast tracts of tropical rainforest that straddle the equator; in the eastern part of this zone are the large lakes and reed beds of the wetlands. South of the equator are Africa's tropical grasslands – the savannas – an immense area of level land dotted with trees. In the veld of southern Africa, the grasslands become subtropical and steppe-like before merging into the dry habitats of the Namib and Kalahari deserts in the far south.

DESERT (PP.40–43)
The Sahara is approximately equal in area to Europe. Most of it is stone and gravel; the rest is made up of sand-dune regions, transition zones, mountainous areas, and a few oases. Desert mammals have highly specialized adaptations to enable them to cope with fierce heat, sparse vegetation, lack of shelter, and, of course, shortage of moisture.

KEY TO HABITATS

- GRASSLAND
- DESERT
- MOUNTAINS
- WETLANDS
- FOREST

MOUNTAINS (PP.44–47)
Africa lacks a "backbone" range of mountains – it does not have the equivalent of the Rockies, Andes, Himalayas, or Alps. The largest high-altitude area is the Ethiopian Highlands, where several scarce species, such as gelada baboons, are found. The Atlas range in the northwest has many animals in common with southwest Europe.

Atlas Mountains
Canary Islands
Erg Iguîdi
Grand Erg Occidental
Grand Erg Oriental
Erg Chech
Ahagga
S a h
Taoudenni Basin
Senegal
S a h e l
Niger
Niger
Cape Verde Islands
Lake Volta
Grain Coast
Ivory Coast
Gold Coast
Slave Coast
Bight of Benin
Niger Delta
Gulf of Guin
São Tomé
ATLANTIC OCE

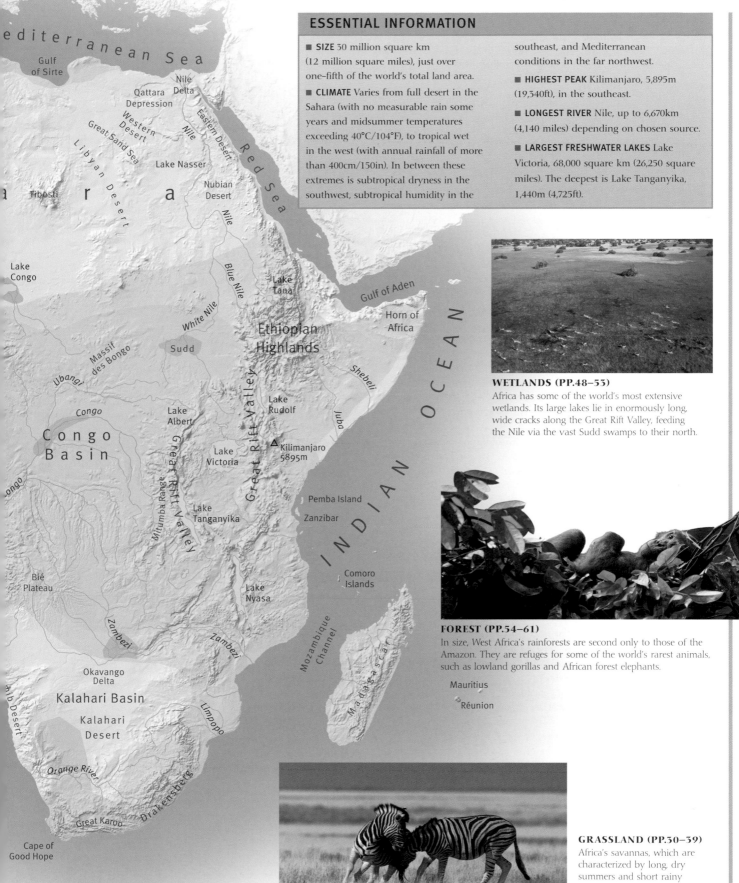

ESSENTIAL INFORMATION

■ **SIZE** 30 million square km (12 million square miles), just over one–fifth of the world's total land area.

■ **CLIMATE** Varies from full desert in the Sahara (with no measurable rain some years and midsummer temperatures exceeding 40°C/104°F), to tropical wet in the west (with annual rainfall of more than 400cm/150in). In between these extremes is subtropical dryness in the southwest, subtropical humidity in the southeast, and Mediterranean conditions in the far northwest.

■ **HIGHEST PEAK** Kilimanjaro, 5,895m (19,340ft), in the southeast.

■ **LONGEST RIVER** Nile, up to 6,670km (4,140 miles) depending on chosen source.

■ **LARGEST FRESHWATER LAKES** Lake Victoria, 68,000 square km (26,250 square miles). The deepest is Lake Tanganyika, 1,440m (4,725ft).

WETLANDS (PP.48–53)
Africa has some of the world's most extensive wetlands. Its large lakes lie in enormously long, wide cracks along the Great Rift Valley, feeding the Nile via the vast Sudd swamps to their north.

FOREST (PP.54–61)
In size, West Africa's rainforests are second only to those of the Amazon. They are refuges for some of the world's rarest animals, such as lowland gorillas and African forest elephants.

GRASSLAND (PP.30–39)
Africa's savannas, which are characterized by long, dry summers and short rainy seasons, are the habitat of great herds of large grazing mammals, such as zebras and wildebeest.

Map labels

Mediterranean Sea
Gulf of Sirte
Nile Delta
Qattara Depression
Western Desert
Great Sand Sea
Libyan Desert
Lake Nasser
Eastern Desert
Nubian Desert
Tibesti
Red Sea
Sahara
Lake Congo
Blue Nile
Lake Tana
Gulf of Aden
Horn of Africa
White Nile
Massif des Bongo
Sudd
Ethiopian Highlands
Shebeli
Ubangi
Juba
Congo
Lake Albert
Lake Rudolf
Kilimanjaro 5895m
Congo Basin
Lake Victoria
Great Rift Valley
INDIAN OCEAN
Mitumba Range
Lake Tanganyika
Pemba Island
Zanzibar
Bié Plateau
Lake Nyasa
Comoro Islands
Zambezi
Mozambique Channel
Madagascar
Okavango Delta
Kalahari Basin
Namib Desert
Kalahari Desert
Limpopo
Mauritius
Réunion
Orange River
Drakensberg
Great Karoo
Cape of Good Hope

DRY-SEASON SAVANNA

OPEN SAVANNA

BUSH SAVANNA

GRASSLAND TYPES
During the dry season, the parched grasses of the savanna wither to a golden brown colour. The arrival of the annual wet season (or seasons) brings a sudden greening and flowering of the grasses and other vegetation. Open savanna is characterized by isolated acacia trees in an ocean of grass; in bush savanna the trees are more numerous, providing food for browsing animals.

GRASSLAND COVERING ABOUT 15 MILLION SQUARE KM (6 MILLION SQUARE MILES), GRASSLAND ACCOUNTS FOR MORE THAN HALF OF AFRICA'S TOTAL LAND SURFACE. DURING THE RAINY SEASON THIS HABITAT CAN SUPPORT VAST NUMBERS OF LARGE GRAZING MAMMALS, AS WELL AS THE CARNIVORES THAT PREY ON THEM.

GREAT GRASSY PLAINS

Most of Africa's great grassy plains are savanna – not pure grassland but a mixture of grass and shrubs or trees. The savanna forms a vast habitat, which stretches across the continent south of the Sahara. The best preserved mammal faunas are in the East African savanna, where the animals roam the famous national parks and other reserves, including the Serengeti, in Tanzania, and Masai Mara, in Kenya.

The proportion of trees in any one area of savanna varies considerably from one region to another. At one extreme is the open savanna: here, the grassland stretches as far as the eye can see, with just a few trees (usually single acacias) or clumps of shrubs scattered across the landscape. At the other extreme is the woodland or forest savanna, where the trees are less densely distributed but form more or less continuous woodland cover. In contrast to the acacias and other shrubs and trees, the trees in wooded and forest savannas are deciduous.

SOMETIMES SOCIAL
Male and female cheetahs generally live apart, seeking each other out only to mate. Once cubs are weaned, adult females are solitary, but males form small groups, which (unlike the females) defend territories.

The intermediate types of savanna include bush savanna, which has small concentrations of trees and bushes, often in the form of thickets. These may contain shrubs bearing vicious thorns (which evolved to deter browsing mammals) and can be so dense that they are impenetrable by humans.

HUGE HERDS

Grasslands in other regions of the world – North America, South America, Europe, and Asia – were also once home to huge herds of grazing mammals. But it is only in the African savannas that so many various species survive in such great numbers. Even so, some have suffered considerable decline as a result of hunting by man; and some, such as the unstriped zebra known as the quagga, have been wiped out altogether.

"…thousands of **chomping** mouths stimulate the **grasses** to produce **fresh growth**."

SAFETY IN NUMBERS
Most grazing animals, such as impalas, live in herds. This makes it difficult for predators to single out a victim.

RENEWING THE FOOD SOURCE

The grazing mammals that depend on the savanna actually help the habitat to renew itself: the thousands of chomping mouths stimulate the grasses to produce fresh growth. Most importantly, this fresh growth contains more nutrients than the grass it replaces; in addition, each plant that has been grazed produces not just a single blade (as in the original growth) but several new blades, each in due course bearing an ear of grain. This further enhances its value to grass-eating mammals because the grass seeds are higher in nutrients than the stems and leaves. This is why grazers – especially pregnant females and mothers that need to produce nutrient-rich milk – often seek out the seed-bearing grasses.

However, the savanna grasses set seed only during particular seasons, when conditions are moist, so grazers must move about to find enough sprouting grass. Many are able to sniff moisture from afar on the breezes so they can locate areas where grass has sprouted.

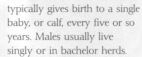

SHARING RESPONSIBILITIES — BEHAVIOUR

Just like humans and other mammals that live a long time, elephants invest a great deal of time and effort in rearing their young. Females and young generally live in extended family groups, under the watchful control of a dominant old female, the matriarch. A female typically gives birth to a single baby, or calf, every five or so years. Males usually live singly or in bachelor herds.

BABY CARERS
The young are cared for not only by their own mothers but by other females in the herd. These carers often attend births in a supportive midwife-like role.

ZEBRA BRAWLS
Rival male zebras push or even bite one another on the neck and legs when competing to mate with females. These fights occasionally result in serious injury.

WALLOWING WARTHOG
Warthogs are extremely fond of lying and rolling about in mud, which helps cool them down and remove parasites.

THE KILL
One female lion rips at a young buffalo's back with her sharp claws while her accomplice tries to get a stranglehold on the victim's neck.

HUNTING FOR A LIVING

In the grassland habitats of Africa, both predators and prey struggle to cope with the seasons, which alternately bring heavy rains and severe drought.

The huge herds of plant–eating mammals that live on the vast expanses of the savanna support a diverse array of predators, from wild dogs and hyenas to the largest of all the carnivores, the lion. Although this impressive cat is the top predator in the savanna, its prey manages to escape more than half of its chases. The other carnivores that hunt on Africa's grassland have a similarly low success rate.

To reduce direct competition, the various species of hunter have evolved to target different kinds and sizes of prey as well as different ways of hunting and killing it. For instance, African wild dogs work in packs, doggedly following zebras, wildebeest, or impalas for great distances until the prey is exhausted and the dogs can co-operate in bringing it down; jackals stalk smaller prey, including livestock, in pairs; servals (p.53) hunt in long grass, pouncing, like a domestic cat, on rodents, birds, and similar small prey; leopards (p.55), masters of solitary ambush, are highly opportunistic and have an exceptionally wide-ranging diet, taking anything from dung beetles and rodents to large antelopes; cheetahs – the world's fastest land animals – run down medium-sized hoofed mammals, such as gazelles and antelope, as well as smaller animals such as hares, in a brief, lightning-quick sprint that can reach speeds of up to 100kph (62mph).

VERSATILE PREDATORS

Lions will hunt in groups, or prides, as well as on their own when they generally restrict their attentions to smaller prey. In general, females do most of the hunting, but males can also be formidable hunters. Lions generally target prey weighing between 50kg (110lb) and 300kg (660lb), from small antelopes to relatively large zebras and wildebeest. But if the conditions are harsh, causing a shortage of suitable

POACHING

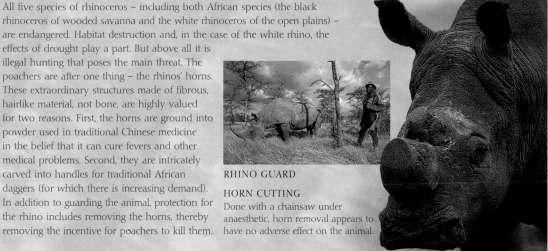

All five species of rhinoceros – including both African species (the black rhinoceros of wooded savanna and the white rhinoceros of the open plains) – are endangered. Habitat destruction and, in the case of the white rhino, the effects of drought play a part. But above all it is illegal hunting that poses the main threat. The poachers are after one thing – the rhinos' horns. These extraordinary structures made of fibrous, hairlike material, not bone, are highly valued for two reasons. First, the horns are ground into powder used in traditional Chinese medicine in the belief that it can cure fevers and other medical problems. Second, they are intricately carved into handles for traditional African daggers (for which there is increasing demand). In addition to guarding the animal, protection for the rhino includes removing the horns, thereby removing the incentive for poachers to kill them.

RHINO GUARD

HORN CUTTING
Done with a chainsaw under anaesthetic, horn removal appears to have no adverse effect on the animal.

large prey, or if a lion is sick or disabled in some way, it will eat rats, small lizards, or even fish. If the occasion presents itself, a pride of lions – or even an individual lion - will attempt to kill larger animals. For example, giraffes, despite their size and their ability to run with huge strides, are occasionally bought down by lions. Male lions, with their greater size and strength, are more likely than females to tackle such huge prey. Such large kills will keep the lion sated with food for a long period.

ARMED AND DANGEROUS PREY

A lion pride may decide it is worth risking an attack on a mammal as large and fierce as an African buffalo (see p.49). Renowned for being unpredictable and aggressive, the buffalo is among the most dangerous of all animals in Africa. A mature male buffalo may stand up to 1.7m (5½ft) high at the shoulders, weigh up to 850kg (1,900lb), and can run as fast as 48kph (30 mph). Moreover, both males and females are armed with a pair of impressive, strongly curved, sharp-tipped horns, which are used to defend themselves and their calves against predators.

Despite their curved tusks and their habit of disappearing into deep burrows to escape both the heat and predators, warthogs (see opposite) sometimes fall prey to lions, particularly if taken unawares when enjoying a wallow in mud. The warthog is the only pig fully adapted to grazing grassland (others forage in forest and woodland) and, like most grassland animals, it has an acute sense of smell and of hearing, and an ability to run fast – up to 55kph (34mph). It is not the prominent upper tusks that lions have to watch out for most when launching an attack on a warthog but the much sharper lower ones, which can cause major wounds.

Other well-protected prey that is nevertheless occasionally tackled by lions are the two species of African crested porcupine. The great cats must be very careful to avoid being injured by the porcupine's long quills. When attacked it will raise its quills and charge backwards towards its tormentor. If the quills make contact, they can cause wounds that often fester and turn septic, sometimes with fatal results.

DANGEROUS PREY
A pair of lionesses cautiously approach a crested porcupine. Its long, needle-sharp spines cannot be shot out, but they can become detached and embedded in the skin.

GRASSLAND DWELLERS

BATS that are at home in dry, barren habitats (of which there are only a few) include Hemprich's long-eared bat of North Africa, which hunts for flying insects.

MOST MONGOOSES are solitary, but the yellow mongoose, found in southern Africa, lives in highly organized packs in which all members share in the care of the young.

ZORILLAS (also called African, or striped, polecats), use their clawed feet to dig for mice and other small prey. When threatened, it sprays noxious fluid from the anal glands.

GIANT AFRICAN mole-rats derive their name from their mole-like lifestyle. They have blunt, rounded heads, adapted for burrowing, and large incisors for gnawing roots.

"The African savanna forms a vast habitat, which stretches across the continent south of the Sahara."

ON THE LOOKOUT FOR TROUBLE
Meerkats – members of the mongoose family – have a highly developed social system in which some individuals look out for predators while others hunt or babysit the young (kits).

TREE–EATERS

By far the tallest of all the mammals living on the African savanna – indeed, the tallest of all animals living in the world today – is the giraffe. Standing up to 5.8m (19ft) tall (females are shorter and considerably lighter) on its great stilt-like legs, and equipped with a neck that accounts for almost half its total height, this extraordinary creature can reach the highest branches of the trees that dot the great expanses of the African grassland. This gives the giraffe a great advantage over the other browsing mammals, which must find their food nearer the ground. The notable exception is the elephant (p.59), whose great strength enables it to reach high foliage by another means:

GERENUK
It cannot compete directly with a giraffe, but by standing on its hind legs and stretching its long neck the gerenuk can access leaves up to 2.5m (8½ft) high, which is beyond the reach of other antelopes.

COAT PATTERNS
Giraffes are divided into six subspecies according to the pattern of their coat. Patterns range from sharply defined, regular patches to relatively indistinct patches. Colouring can also vary from deep chestnut or yellow to almost black.

MATERNAL CARE
A giraffe cow can produce five to ten young (or calves) in her lifetime, one at a time. Twins are very rare.

it simply uses its massive head and trunk as ramming tools to knock the tree down.

To maintain their huge frames, giraffes need to spend much of their time feeding. They are especially fond of the leaves, twigs, and flowers of the various species of acacia tree. Acacias protect themselves against browsers with viciously sharp thorns up to 5cm (2in) long, but this is no defence against the determined and well-equipped giraffe. The animal wraps its long, tough tongue – which can reach up to 53cm (20in) – around the thorny shoots and plucks them wholesale. Powerful molar teeth in the rear of the jaws then crush the prickly food so that it can be swallowed. Other adaptations, including thick, very sticky saliva and a highly specialized liver, enable giraffes to cope with this tough diet, which also contains chemicals highly toxic to other animals.

The males are larger than the females, so they need to eat more, which perhaps explains why they appear to prefer areas of savanna that are more densely wooded with taller trees as these will supply them with more food in a given time. The pattern of dark blotches on the giraffe's coat is especially useful in this environment because it helps to camouflage them against the background of dappled light and shade of wooded savanna. Females tend to live in more open areas that have shorter, more scattered trees: this is especially true when they have young with them, when the clear view enables them to spot approaching predators from far off.

RUNNING FROM DANGER

Giraffes have few predators apart from humans and lions. Although they appear to be moving slowly as they sail across the savanna, in reality they travel at an impressive rate of over 50kph (30mph), with each stride of those great legs covering between 3m (11ft) and 5m (16ft) of ground. They can therefore usually outrun most predators, provided they notice their enemy approaching and are on level ground. Even when cornered, a giraffe can be a terrifying adversary, kicking out with its long, powerful legs tipped with hard hooves.

However, if a giraffe is caught unawares, or is driven on to rough, broken terrain and so forced to slow down or stumble, then even a fully grown adult bull giraffe may be no match for a determined pride of lions. Indeed, in some areas, such as the Kruger National Park, researchers have found that almost twice as many male giraffes as females are killed by lions. This may be because the males spend more time in the more densely wooded areas where they are easier to stalk, or it may be that the heavier males are slower at escaping.

INDEPENDENT RESPONSES
Giraffes usually live in small herds. Unlike the other large herbivores, which live in densely packed herds, giraffes move individually – not as one – at the approach of a predator.

REACHING UP

EVOLUTION

Modern giraffes have necks that measure over 2m (6½ft) long. But the giraffids that lived in Africa millions of years ago – at a time when great rainforests spread across much of the entire continent – had much more modest-sized ones. Surrounded by leafy plants at low levels, they were able to find all the food they needed. Gradually, as the climate changed, the forests dwindled and the trees became more widely scattered. Giraffids that had longer necks were able to survive because they could reach leaves near the tops of the trees, avoiding competition for an increasingly scarce resource with other browsing mammals.

DRINKING
To drink, a giraffe must adopt this splayed posture, otherwise the rush of blood from heart to brain could be fatal.

Legs grew longer to suit life on svannah

Stripey camouflage suited to forest evolved into patches suited to dappled shade

HELLADOTHERIUM
In comparison to the modern giraffe, this prehistoric animal had a short neck and short legs.

PALAEOTRAGUS
The longer neck of this okapi-like giraffid gave it an advantage over other browsing mammals.

LONG NECK
Surprisingly, the neck of the modern giraffe contains the same number of vertebrae (seven) as a human's.

WILDEBEEST ON THE MOVE
Wildebeest behave as flock: if one moves the whole herd moves. When running from danger they can reach speeds of up to 80kph (50mph) – about the same speed as a lion.

ANTELOPES

Africa's grasslands support a range of herbivores collectively known as antelopes. The grazing antelopes can be divided into three main groups, called tribes: the Hippotragini, or horse-like antelopes, such as the roan antelope, found in dry savanna; the Alcelaphini, including the hartebeest, found in open woodland and moist savanna; and the Reduncini, such as the bohor reedbuck and puku, found in tall or tussocky grassland and wetland. Some of the dwarf antelopes (Neotragini), which include the oribi, live on the grassy plains.

ROAN ANTELOPE

HARTEBEEST

ORIBI

BOHOR REEDBUCK

PUKU

MASS MIGRATION

Each year, the grasslands of East Africa are witness to one of the earth's most dramatic of all wildlife spectacles – the great circular migration of grazing mammals in response to the changing wet and dry seasons. Wildebeest, in particular, form vast herds containing up to half a million of these wispy-haired, shaggy-bearded, curved-horned antelopes. Sometimes known by their alternative name of gnu (from their strange grunting calls), wildebeest often travel in mixed herds that include another antelope species, the topi (p.49), as well as zebras (p.31), and various species of gazelle.

As the drought advances, these grazers must stream across the savanna to find new areas of croppable grasses or they will face starvation. Each year they travel more than 1,300km (900 miles) across the plains, crossing rivers and braving dangers from predators – crocodiles, lions, African wild dogs, and spotted hyenas – as well as from hunger and thirst. There is no time to make long, leisurely stops for breeding: calves, born on the move and up on their feet within minutes of birth, must keep up with the herd within days or succumb.

TERMITE DIET

Among the most conspicuous features of the African grasslands are the huge, rock-like edifices constructed by termite colonies. Through their ability to digest tough cellulose and aerate the soil, these superficially ant-like creatures are essential to the savanna's ecosystem. They are also highly nutritious food for a range of animals, from jackals, warthogs, and many birds, to specialist termite-eating mammals such as aardvarks, aardwolves, and pangolins.

A PATIENT TERMITE PREDATOR
The aardwolf (right) stands patiently outside a termite mound (inset). Rather than breaking into the their homes, it waits for the termites to emerge so that it can then lick them up.

PANGOLIN
The pangolin's pinecone-like armour protects its body from the bites of soldier termites as well as from predators.

The aardvark (see right) is a solitary, nocturnal animal that sleeps in temporary burrows during the day. It has poor eyesight but an excellent sense of smell, which it uses to locate ants or termites. It then breaks open the insects' home and probes the tunnels with its long, sticky tongue. The pangolin (see left) operates in a similar way to the aardwark. The aardwolf (see below) is also solitary and nocturnal. A relative of the hyena, its diet of termites is supplemented by grubs and other soft-bodied creatures.

Chimpanzees (p.56) occasionally eat termites, and some chimp communities display an ingenious way of accessing termite mounds. They select a suitable stick or blade of grass and insert it into a tunnel. The termites attack the stick, clinging on with their mandibles. The chimpanzee then withdraws the stick and licks it clean.

AARDVARK
Using its huge, curved claws and powerful muscles, the aardvark rips into termite mounds. It can eat the contents within 30 seconds.

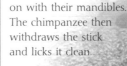

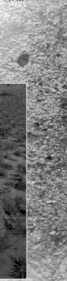

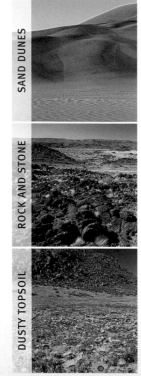

SAND DUNES

ROCK AND STONE

DUSTY TOPSOIL

DESERT TYPES
African deserts contain vast tracts of sand swept up into mobile dunes by the wind. But they also have rocky and stony areas, which sometimes retain enough dusty topsoil to support annual plants that burst into flower after rainstorms.

DESERT COVERING A VAST AREA OF AFRICA, THE ARID, SUN-SCORCHED TERRAIN OF THE DESERT MAY SEEM TOO BARREN TO SUSTAIN ANY KIND OF LIFE, BUT IT IS HOME TO A WIDE VARIETY OF MAMMALS THAT HAVE EVOLVED ADAPTATIONS FOR SURVIVAL IN ONE OF THE MOST HOSTILE ENVIRONMENTS ON EARTH.

THE DRY CONTINENT
Much of the land to the north and south of Africa's forest zone is desert. In the north lies the Sahara, the largest desert in the world. It has a total area of over 9 million square km (3½ million square miles), and extends across 14 north African countries. At the heart of southern Africa lies the Kalahari, which covers 570,000 square km (220,000 square miles). The Atlantic coast of southwest Africa is fringed by the Namib Desert, which is 2,000km (1,240 miles) long and covers nearly 34,000 square km (13,000 square miles).

Some areas of these African deserts are great seas of wind-blown sand, but others are forbidding landscapes of sand-blasted bare rock and gravel. They all share one feature that is hostile to life: drought. Over the year, there is not enough rainfall to offset the drying effects of the sun. As a result, the land cannot

DESERT SPECIALIST
The addax lives in the desert of northwest Africa. It is able to get all the moisture it needs from its food.

support dense vegetation, so herbivorous animals spend much of their time searching for the few plants that do grow. Some plants are adapted to resist drought by conserving every drop of moisture that is available to them. Others sprout, flower, and set seed in the brief periods that follow the infrequent desert rainstorms; the seeds then lie dormant, often for years, until the next rainstorm starts them into life.

The mammals that live in the desert have two main concerns: they must avoid drying out, and they have to find enough food.

DESERT NOMADS
Because plant life is sparse, large grazing mammals – such as camels, desert antelopes, and wild asses – are nomadic, wandering over vast areas to find sufficient food. They are well adapted to eat the tough

PORTABLE SUN SHADE

Most small desert mammals are strictly nocturnal, but the African (or Cape) ground squirrel of southern Africa is active by day. This is possible because unlike most ground squirrels it has a bushy tail, which it uses as a sun shade, arching it over its back to help protect itself from the desert sun. African ground squirrels live in single-sex colonies, males and females coming together briefly to breed before separating again.

OPPORTUNISTIC FEEDERS
African ground squirrels feed on roots, leaves, seeds – whatever they can find near the safety of their burrows.

foliage of the desert perennials, and they have a talent for locating the flushes of vegetation that sprout in the wake of rainstorms. But they are too large to shelter underground or beneath rocks by day, so they need special adaptations to help them cope with constant exposure to heat. Most have light-coloured fur to help reflect heat, and many have long legs, which not only allow them to cover long distances but keep the body away from the scorching sand or soil.

Water conservation is essential. One tactic is to drink huge quantities when it is available and store it in the body for use during drought. Camels drink up to 135 litres (30 gallons) at once and store it in their stomachs. But camels and other mammals are also adapted to avoid losing too much moisture. They produce very little urine, and a camel can allow its body temperature to rise well above 40°C (104°F) before it starts to sweat. Camels also have slit-like nostrils, which, as well as protecting the nasal passages in sandstorms, reduce the moisture that would otherwise be lost in the air they breathe out.

DESERT DWELLERS

MOUSE-TAILED bats fly only by night. They store fat to enable them to survive periods when food is scarce.

CATS, such as this black-footed cat, hunt in the dark, seeking out the small mammals that feed at night.

JACKALS are highly adaptable, opportunistic omnivores – they will eat almost anything.

FAT-TAILED gerbils and many other small mammals nibble seeds and grasses by night.

MOLES have almost invisible eyes and ears and rarely surface above the ground.

"Mammals that live in desert have two main concerns: they must avoid drying out, and they have to find enough food."

SUPERBLY ADAPTED FOR THE DESERT
Camels have long legs with splayed, padded feet that enable them to walk easily on sand. They can go for weeks without food and water, living on the fat stored in their humps.

COOL BURROWERS

Unlike the large grazers, mammals such as mice, jerboas, and gerbils can avoid the worst of the heat by hiding in burrows during the day. They emerge at night, when the air is cool, to nibble at sparse vegetation and dormant seeds. They can usually find enough to eat because they have only small appetites. Their need for moisture is met by the dew that covers their food during the night. Like most other desert animals, their digestive systems are adapted to maximize water retention, so they need to drink less than mammals adapted to life in other environments.

DESERT HUNTERS

Despite the apparent scarcity of food, a wide range of animals survive in the desert. Insects, spiders, and scorpions are common, as are snakes and lizards. These coldblooded animals use less energy than warmblooded mammals do, so they can survive on far less food. Such animals provide food for wild cats, foxes, jackals, hyenas, and a variety of other hunters, which also prey on small mammals such as mice and gerbils. Most desert carnivores, and omnivores such as the black-backed jackal (p.41), are opportunistic, so they will also scavenge carrion.

Carnivores typically hunt by night, when the

DESERT GAZELLE
The only gazelle that lives in the deserts of southwest Africa, the springbok even ventures into the arid heart of the Namib. It can survive indefinitely without drinking, provided its food contains at least 10 per cent water.

LONG HOP
A desert jerboa's long legs allow it to range widely in search of food. Its long tail helps it to balance.

(p.41)

SCAVENGING   BEHAVIOUR

Death is never far away in the desert, and dead animals provide a feeding opportunity for scavenging carnivores such as jackals and hyenas. The acid-bath digestion of a hyena (which can process rotten meat, skin, bone, and even teeth), and its exceptionally strong, bone-crushing jaws enable it to make a meal out of carcasses that most other carnivores would reject.

BEACH BONANZA
A dead shark stranded on the coast of the Namib desert makes a feast for a hyena.

smaller animals emerge from their burrows and the desert air is at its coolest. Many have special adaptations that enable them to operate effectively in the dark. The fennec fox (p.16), for example, has huge ears, which help it to detect the slightest scuffle in the night and identify a potential meal; such large ears also act as very efficient heat radiators by day. Many, such as the desert jerboa (see left), also have large eyes.

(p.16)

VITAL RESOURCE
Visiting waterholes on the fringes of the desert is essential for animals that spend much of their time feeding in arid lands. However, during the dry season they can become severely depleted, or even dry up altogether; only the fittest animals survive.

"Despite the apparent scarcity of food, a wide range of animals survive in the desert."

ALERT AND VIGILANT
Like many savanna antelopes, the wary steenbok often moves into desert to feed on the flushes of grass that appear after rainstorms.

AFRO-ALPINE

SUBALPINE

MOUNTAIN TYPES
In Africa's high-altitude zones temperatures can be so cold that mountain peaks are covered in snow. At lower, subalpine altitudes, seasonal rainfall and melt-water from snows on some of Africa's highest peaks join to form the headwaters of often massive river systems.

A WARM COAT
The thick fur of the Barbary ape protects it against the chill air of the Atlas mountains, its last refuge in Africa.

MOUNTAINS

AFRICA'S ATLAS MOUNTAINS IN THE NORTHWEST, THE DRAKENSBERG RANGE IN THE SOUTH, AND THE ETHIOPIAN HIGHLANDS IN THE EAST, PROVIDE REFUGE FOR SOME OF THE WORLD'S RAREST ANIMALS. KILIMANJARO IS ONE OF THE FEW LOCATIONS WHERE ALPINE HABITATS ARE FOUND ON THE EQUATOR.

MOUNTAIN VARIETY

Africa is often thought of as a continent of vast deserts and tropical rainforests, but it also has a number of mountainous regions, where temperatures frequently fall below freezing. The highest elevations – Mt. Kilimanjaro (5,895m/19,340ft) in Tanzania (Africa's highest mountain), Mt. Kirinyaga (5199m/17,057ft) in Kenya, Mt. Elgon (4,321m/14,178ft) on the Kenya–Uganda border, and Mt. Toubkal (4,165m/13,665ft) in Morocco – are snowcapped for most of the year.

Most of Africa's highlands have been created by the uplift of the Earth's surface while subsidence of the surrounding areas has created great basins. Only the Atlas mountains in the northwest and the Cape ranges in the southeast are fold mountains (created by the folding rather than the lifting of the Earth's surface).

The land lying to the south and east of the Congo and Nile basins is, in general, higher than that of North and West Africa. The Great Rift Valley in the east, which stretches almost 4,800km (3,000 miles) from the Red Sea to Mozambique, was created by a mixture of profound faulting and volcanic activity; many of the highest peaks in East Africa are volcanic in origin. South of the Zaire Basin the average elevation of the land is 900m (3,000ft). These plateaux usually take the form of great plains punctuated with occasional rocky outcrops or mesas (table-shaped hills).

MAKING HEAT
Rock hyraxes have poor control over their body temperature so they huddle together for warmth. Crevices provide some refuge from the cool mountain temperatures.

RARE AND DIVERSE HABITATS

The difference in latitude between one mountain range or elevated area and another means that Africa's montane habitats are very varied. For example, the warm climate in the mountains near the equator makes it possible for trees to grow at a much higher altitude than they do at other latitudes, and this creates the special habitat required by some of the world's rarest animals, such as the mountain gorilla (p.56).

Within the range of a mountain's elevation, climate and vegetation vary markedly. This dictates the food resources available and therefore the mammal fauna that each zone of a mountain will support. In the Ethiopian Highlands, which rise at their highest point (Ras Dashen) to 4,620m (15,158ft), the cool, moist climate supports a number of rare species, including the striking gelada baboon (see opposite), which lives in social units of up to 600 individuals between elevations of 2,000m and 5,000m (6,400ft and 16,000ft). The diet of the gelada consists in the greatest part of roots, seeds, and blades of grass. However, the ambient temperature of the Ethiopian Highlands is increasing, and so the lowest elevation at which the montane vegetation will grow is rising. As a result, gelada baboon colonies are finding themselves "marooned" on the tops of the lower peaks without sufficient food resources. Human expansion also diminishes the gelada's habitat, and the splendid mane of the males makes them a target for hunters.

MADE FOR MOUNTAIN LIFE

Many of the mammals that live in the different mountain zones have evolved unique behavioural, physiological, or morphological adaptations that equip them to cope with mountain climate and terrain. For example, the insulating properties of their coats are high in

---

SURE-FOOTED CLIMBERS — EVOLUTION

Powerful limbs enable the Barbary sheep to leap up to 2m (6½ft). Like other bovids adapted to mountain living, it also has small, hard hooves for negotiating rocky ledges, and under its feet it has non-slip pads to provide good grip. Most land mammals are afraid of steep drops, but these mountain-climbing mammals have a strong sense of balance and no fear of heights. Hunting, and the expansion of human habitation and domestic cattle herds, has resulted in loss of habitat for the Barbary sheep as well as for species such as the Nubian and Walia ibex. In Morocco, Barbary sheep can be found as high as the snow-line at 3,800m (12,500ft) while the Walia ibex will range as high as 4,500m (14,700ft).

REMOTE REFUGE
The Barbary sheep favours remote areas that are not easily accessible to hunters.

"The tropical **mountains** of **Africa** provide habitats for some of the **world's** rarest animals".

comparison to those of lowland mammals, and those that live in rocky uplands are good climbers. Some of the best are the hoofed mammals, such as the Barbary sheep (see box, opposite) and the ibex of North Africa. The rock hyrax (see opposite) is also found in rocky uplands, where water and vegetation can be sparse. To cope with this, its kidneys are remarkably efficient, so it requires very little water and it will eat almost every type of plant.

Above the treeline, where the land is exposed to hot sun during the day but temperatures plummet at night, many small mammals are burrowers, excavating deep tunnel systems that provide refuge from the conditions above ground. At high altitudes, the oxygen content of the air is low, so mammals that live permanently at high altitude are physiologically adapted to make exceptionally efficient use of the oxygen in thin air.

FACIAL GLAND BEHAVIOUR

A peculiarity of some bovids, such as the royal antelope, the long-snouted dik-dik, and the klipspringer (right), is the presence of a noticeable facial gland positioned immediately in front of the eye. Secretions from the gland are used by both females and males to scent-mark plant stems and branches of trees. The hardened secretion around the gland will be nibbled by both sexes during grooming sessions. The klipspringer is almost extinct outside the national parks within its range.

ON TIPTOES
The klipspringer is one of the most sure-footed of mountain dwellers. Its tiny feet are so much more upright than most bovids that it appears to be moving on tips of its hooves.

ETHIOPIAN HOMELANDS
The rare and endangered gelada baboon lives in large troops in the rocky highlands of the Ethiopian massif. Both sexes have a bright red patch in the middle of the breast.

LIVING IN THE UPLANDS

The variety of vegetation found at different altitudes on the African mountainsides supports many large mammals and a great number of small ones, mainly rodents such as rats and mice. One of the most recognizable is the giant African mole-rat (see box, opposite), which inhabits the mountainous regions and grasslands of East Africa. The loose soil of the volcanic uplands is ideal for its burrowing activities.

One of Africa's larger mountain mammals – and one of the most endangered – is the Cape mountain zebra (see opposite), distinguishable from its lowland counterparts by the orange marking above the muzzle. It is the smallest of the zebra family, and it is also the most geographically restricted zebra. It formerly inhabited all the mountain ranges of the south and southwest, but is now reduced to tiny populations in national parks.

CAUGHT
More than 30 African wild dogs may join in the hunt. By outnumbering their prey and co-operating in this way they increase the chances of making a kill.

RARE DOGS

The African wild dog (sometimes called the Cape hunting dog) is an elusive and rarely seen predator. Its scientific name, *Lycaon pictus*, meaning "painted wolf", is derived from the variable swirls and patches of black, white, and yellow that make up the animal's patterned coat coloration. Its huge skull, short hair, and powerful jaws suggest a similarity to hyenas; it has also been likened to wolves, although wolves are much heavier and have shorter legs. However, its movement and whining vocal expression link it more closely to certain breeds of domestic dog. Its muscular body and long limbs are ideally suited to rapid locomotion (it can run up to 48kph/30mph), but for its success as a hunter the African wild dog relies mostly on its remarkable stamina, which allows it to pursue powerful animals, such as wildebeest and zebra, over great distances.

PACK SPOILS
Having tirelessly pursued their prey to exhaustion, African wild dogs gorge themselves on the meat. Some of this meal will then be regurgitated for the young.

It may be ferocious and savage when hunting, but with its own kind the African wild dog is one of the most social of the canids. Packs are highly organized, and individuals are responsive and playful towards each other. Packs numbering between one and three dozen individuals will live contentedly together and will raise litters communally. They hunt co-operatively, and pups are allowed to eat before all other members of the pack, including the dominant pair. Bitches ordinarily give birth to between six and ten offspring, although as many as 16 pups have been reported in a single litter.

These dogs are constantly on the move in search of food, but as soon as pups are born the parents will seek a safe refuge for their

SIMIEN FOX
One of the rarest and most elusive members of the dog family, the Simien fox can now be glimpsed only in two small national parks in Ethiopia.

young, often making use of an abandoned burrow made by another animal such as a warthog or aardvark – and the pack will then remain in this place until the pups are old enough to travel.

African wild dogs were once common and widely distributed in all habitats except rainforest and full desert. Now they are one of Africa's most endangerd carnivores, existing in fragmented populations in savanna and in mountainous tracts such as those of Kilimanjaro. The isolated massif of the Ethiopian highlands provides a refuge for the critically endangered Simien fox (see above). Sometimes called the Ethiopian or Abyssinian wolf, or the Simien jackal, this rare member of the dog family exists only in small numbers within the protected confines of the Simien and Bale mountains. It is highly unusual among canids because its diet consists primarily of rodents such as hares and various types of rat, including the giant African mole-rat (see right). This diet has enabled it to survive at high altitudes where rodent populations are high.

UPLAND BABOONS

At lower altitudes (up to 2,000m/6,400ft) in Ethiopia and throughout the Horn of Africa, the hamadryas baboon inhabits rocky, sparsely vegetated outcrops. These baboons live in close family units consisting of a dominant male and up to ten females and their joint offspring. When travelling and foraging, the family will join other families to form a large band. This social system is not adopted by the highly intelligent and adaptable chacma baboon (see right), which can be found near water in both the upland and coastal areas of southern and southwestern Africa. Chacma baboons do not live in family units but in large troops consisting of a dominant male and female pair, and other adults and offspring.

MOUNTAIN DWELLERS

MOUNTAIN ZEBRAS are generally divided into two subspecies: the Cape mountain zebra (shown here) and Hartmann's mountain zebra.

BABOONS are highly adaptable omnivores. The chacma baboon will eat anything, from a young gazelle to insects and fruit.

HYRAXES are highly skilled mountanineers. Glandular secretions on their footpads help their feet to adhere to steep rockfaces.

BURROWING RODENTS include the giant African mole-rat, which digs extensive underground tunnels.

> "...**ferocious** and savage when hunting, with its own kind the African **wild dog** is one of the most social of the **canids**."

WETLAND TYPES
Many African wetlands
have vast expanses of reeds
or papyrus, which shelter
a wide range of wetland
mammals. Some wetlands
are broad expanses of open
water that provide homes
for true water mammals
such as the hippopotamus.

WETLANDS
THE RICHNESS OF WETLAND
HABITATS RIVALS THAT OF RAINFORESTS. BUT IN THE TROPICAL
AFRICAN CLIMATE THEY CAN ALSO BE UNSTABLE, CHANGING QUITE
DRAMATICALLY WITH THE SEASONS, SO THE MAMMALS THAT LIVE
IN THEM MUST BE ADAPTABLE.

GREEN OASIS
The rivers and lakes of Africa are
fringed by extensive wetlands that
provide water, food, and refuge for
a wide variety of mammals. Many
of these wetlands are oases of lush
vegetation surrounded by arid land.
In Botswana, for example, the
Okavango river drains into the vast
soakaway of the Kalahari desert to
create the great swamp known as
the Okavango Delta. In southern
Sudan the headwaters of the White Nile spread out on
the southeast fringes of the Sahara to form the Sudd,
a huge expanse of papyrus swamp that covers 130,000

SWAMP ANTELOPE
Sitatunga are wetland specialists, which often
wade shoulder-deep through the water while
feeding on reeds and rushes.

square km (50,000 square miles) – the
largest freshwater wetland in the world.
And throughout the continent's deserts
and savannas, smaller wetlands
around rivers and pools support
their own wildlife communities.
All these wetlands ebb and flow
with the seasons. During the dry
season the Okavango Delta shrinks
from over 12,000 square km (4,600
square miles) to 4,000 square km
(1,540 square miles), and smaller
swamps and waterholes may dry up altogether. So
although, for the animals that depend on them, wetlands
are rich habitats, they are also extremely challenging.

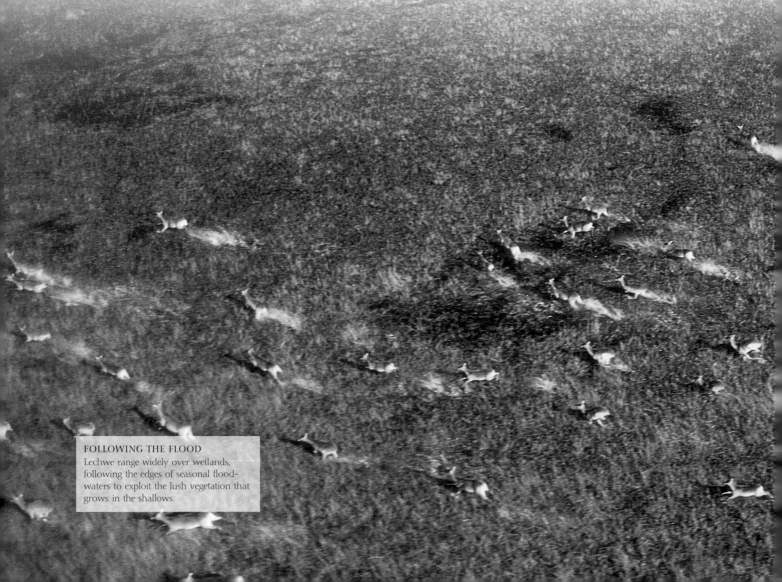

FOLLOWING THE FLOOD
Lechwe range widely over wetlands,
following the edges of seasonal flood-
waters to exploit the lush vegetation that
grows in the shallows.

INSECT REPELLENT

Many large African mammals regularly wallow in water or mud. They do this for a number of reasons. Mud is a good coolant because the water in it evaporates more slowly than clean water, and caked mud makes a very effective sunscreen. It also acts as a barrier against bloodsucking flies and ticks, and since these may carry diseases – some of them fatal – the mud could be considered a lifesaver.

MUDPACK
Caked in mud, an African buffalo makes the most of nature's coolant, sunscreen, and insect repellant.

VISITORS AND RESIDENTS

African wetlands attract many different kinds of mammal. These include several species that are not specifically adapted to the water, such as the zebra, wildebeest (p.38), and topi, which visit to drink each day, and the elephants, rhinos, and African buffalo (see left) that come both to drink and to bathe. Elephants in particular enjoy rolling and wallowing in water, both to cool themselves and to wash away parasites. If it is deep enough they may submerge themselves completely; in shallow pools they bathe by drawing water up into their trunks and spraying their bodies.

Other large mammals found on wetlands are residents rather than visitors, with adaptations for life on soft, wet ground. They include the wetland antelopes of the tribe Reduncini (p.39), such as the reedbucks, waterbuck, kob, and lechwe (see box, right). These antelopes are not restricted to wetlands, but they depend on the lush vegetation that grows on moist grasslands and are never found far from water. Waterbuck and lechwe often wade into deep water to eat aquatic plants, so they have feet that splay widely to bear their weight on the soft mud. This feature is shared by the sitatunga (see opposite), a spiral-horned antelope that lives exclusively in the papyrus swamps and other deep marshlands of south-central Africa.

ANTELOPES

Various antelopes are adapted to living in or near seasonally flooded plains or swamps. They will wade or swim both to find food and to escape predators. The waterbuck in particular will use the water as a hiding place, submerging all but its nose.

WATERBUCK

LECHWE

TOPI

VALET SERVICE
When swimming under the water, hippos often attract fish that nibble algae from their skin and even from around their teeth.

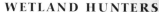

PYGMY HIPPO
Smaller and rarer than the common species, and adapted to spending more time on land, the pygmy hippopotamus lives by rivers and swamps in the lowland forests of west Africa.

WATER BABY
Hippo calves are usually born under the water. They will also suckle while submerged, surfacing every few seconds to take a breath of air.

BATHING COMPANIONS
Even though they do not live in structured herds, hippos are quite tolerant of crowding in the water. During the dry season, when pools shrink, large gatherings may form.

WETLAND HUNTERS

The many plant-eating animals that feed on the lush wetland vegetation attract a variety of hunters. The serval – a lithe, long-legged cat – specializes in hunting among long grass and reeds. It pounces on grass rats and other rodents, but it also hunts through the shallows for frogs, fish, and even flamingoes. Flamingoes are also targeted by jackals and hyenas, which prowl around the fringes of the great flamingo flocks that live on the lakes of the African Rift Valley. Any bird that strays from the flock is likely to fall victim to a dash-and-grab attack.

Although they are often taken by crocodiles, the wetland antelopes that favour deeper water are usually fairly secure from attack by other mammals. But the herds of antelopes and zebras that visit the wetland fringes in search of water and good grazing are shadowed by hunters such as lions (p.32), leopards (p.55), and painted hunting dogs. During the dry season, especially, the water edges attract large crowds of grazing animals in search of the surviving green grass. This is a lean time for them, but for many of their predators these conditions make for easy hunting: with so many animals gathered in one place, they are spoilt for choice. A pack of dogs, for example, will simply charge a herd to get the animals running, then pick off the weaker animals that lag behind.

AQUATIC GIANT

Few predators ever attempt to tackle the most aquatic of all African mammals, the hippopotamus. Growing to a weight of 3,000kg (6,600lb) or more, and armed with sharp tusks, a hippo is a formidable animal. Specialized for life in the water, it has webbed toes, and its eyes, ears, and nostrils are sited on the top of its head so it can be almost completely submerged while still being able to breathe. Indeed it must spend most of its time in the water because it has a uniquely thin skin that loses body moisture very quickly; if it stayed on land under the tropical sun the skin would soon

crack and the animal would become dehydrated and die. It gains some protection from a red fluid that oozes from mucous glands in the skin and dries like lacquer (giving the animal its pinkish skin colouring). However, the only way it can cool down and prevent itself from drying out is to plunge below the water.

HIPPO TRAILS

Hippos spend the day wallowing in and under the water, but at night, when the air is cooler, they emerge to feed on dry land. Each hippo eats about 40kg (90lb) of grass a night, which seems a lot although it is only a small percentage of its body weight. Hippos favour grass that is kept short by regular grazing, so they often return to the same feeding areas night after night. This regular traffic creates well-defined trails. In the swamps of the Okavango Delta, the resident hippos make channels through the swamp that lead to grassy

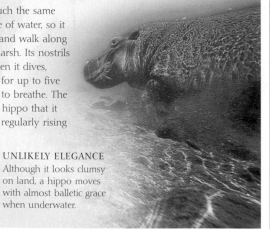

INSTINCTIVE DIVER

A hippopotamus's body has much the same weight as the equivalent volume of water, so it can submerge like a submarine and walk along the bottom of a river, lake, or marsh. Its nostrils and ears close automatically when it dives, and it can stay under the water for up to five minutes before it has to surface to breathe. The process comes so naturally to a hippo that it can even sleep under the water, regularly rising to the surface, breathing, and submerging itself again without waking up. It may spend most of the day like this, before leaving the water to graze at dusk.

UNLIKELY ELEGANCE
Although it looks clumsy on land, a hippo moves with almost balletic grace when underwater.

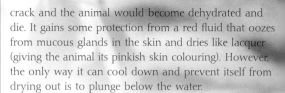

LEOPAR
powerful
on the sa
quarry up
they can

a close an
lighter tha
climber, th
along bou

ON TH
One of the
forest is th
canines gr
up to 20, it
bamboo fr
areas of Et
family, the
mostly lush
plunder foo

Althoug
pouched ra
allowing th
the animal'

LONE HUNTER
The stealthy and secretive serval hunts
alone, using its large ears to locate the
sound of moving prey among the grass
and wetland vegetation.

islands where they can feed. These channels guide the
flow of floodwater during the rainy season, so the
hippos have actually defined the form of the wetlands.

Each hippo forages alone for food. The adult does
not need to worry about predators, so there is no
advantage in staying together. Yet despite this, females
will gather in tight groups. Mature males, on the other
hand, are not sociable. They will drive rivals away, if
necessary fighting with their sharp lower tusks. If one
of the males does not give way, these fights can be
extremely bloody and occasionally the loser is killed.

"In the dry season, the water edges
attract large crowds of grazing
animals in search of the surviving
green grass."

53

RAINFOREST

The African rainforest is divided roughly into two great belts of land. The first, and the smaller, area is the Upper Guinea forest, which occupies lower areas from Sierra Leone on the continent's Atlantic seaboard to western parts of Ghana. Here can be found Diana and mona monkeys, duikers, civets, porcupines, and pangolins (p.54). In dense rainforest, many animals rely on sound rather than sight to communicate with each other. The most vocal are the black and white colobus monkeys, which make high-pitched howls and screeches at dusk and dawn.

The second, and larger, area of rainforest stretches from the northern coastline of the Gulf of Guinea across Cameroon, Gabon, and the Congo Basin to the Ruwenzori mountains on the border with Uganda. This is the "jungle" often associated with Africa's interior: a humid, densely vegetated environment that supports three species of a rare and well-studied primate, the gorilla. The eastern lowland gorilla can be found only in the forests of eastern Congo; the western lowland gorilla is more widely distributed throughout the region; the mountain gorilla (see above) is found at higher elevations and occupies only a few small enclaves in the eastern part of the range.

LEARNING
By staying close to their mothers, young gorillas gain first-hand experience of how to deal with forest life.

MOUNTAIN FORESTS

The montane cloud forest that clings to the high slopes of the volcanic Virunga range of mountains between Uganda, Rwanda, and the Congo is home to the only known populations of mountain gorilla. Each group is dominated by an alpha male, who will robustly protect both his offspring and the harem of females that produced them.

Gorillas employ over 20 separate vocalizations, but they also use body language to communicate both with members of their group and with outsiders. When repelling the advances of an unwanted intruder, including humans, a dominant silverback follows an established pattern: he will emit a number of cries before holding a plucked leaf between his lips; after a series of more urgent cries he will stand upright while hurling leaves and twigs into the air. He will then beat his chest and run sideways while ripping up vegetation. As a final act of repulsion, the great primate will beat the earth and may then charge.

THREATENING BEHAVIOUR
A silverback will keep group members in check and warn off intruders by standing erect and beating his massive chest.

DWINDLING NUMBERS

Gorillas are peaceable by nature: generally they attack only when they perceive a direct threat or in order to protect their family group (see above). They inhabit often remote tracts of land – about 25–40 square km (9–15 square miles) per group – and have few natural predators, yet their numbers are declining.

Other than the occasional attack by a leopard, the two main threats to their existence are occasioned by humans. Their hands, feet, and even their heads, are sought after by collectors, and this has given rise to a poaching industry. The second threat comes from increasing agricultural expansion, resulting in the clearing of significant tracts of forest and, therefore, much of the gorillas' typical habitat. Although national parks with armed patrols have been established to protect gorillas, poaching remains one of the most significant threats to the species' survival.

GROOMING CHIMPS
Chimps spend many hours in grooming sessions, which serve not only to keep a chimpanzee's fur free from parasites but to create and strengthen social bonds between the different members of the group.

The gorilla is the largest of all primates: adult males weigh an average of 160kg (350lb) and, if they were to stand upright with straightened legs, would reach a height of 2.5m (7½ft). Baby gorillas adapt to their environment quickly and develop much faster than human offspring. By the time they are 6 or 7 months old, they are strong enough to climb trees. Mountain gorillas have adapted to life at altitudes of up to 4,000m (13,200ft) by growing long, shaggy hair to keep them warm

SILVERBACK
When adult male gorillas reach sexual maturity at 9 or 10 years, their backs turn silvery grey. The "silverback" goes on to become the dominant group member.

FAMILY AT REST
A gorilla group, comprising several young females and juveniles, relaxes in the heat of the day. Young gorillas often play while the adults sleep.

MANDRILL PORTRAIT
The facial coloration of the adult male mandrill denotes his authority and wards off enemies. The red and blue markings are repeated on the hindquarters.

IN CLEARINGS

The gradual recession of the forest boundaries during the Pliocene period (5.3 to 1.8 million years ago) forced many creatures, such as baboons, from their natural arboreal habitat onto the plains, there to adapt to a terrestrial lifestyle where their tree-climbing abilities were of little use. Some groups of mammals eventually returned to the forests, not to inhabit the canopy but to utilize the natural cover and food resources of the forest floor. Two such species, the mandrill (see opposite) and its smaller relative, the drill, can occasionally be observed in the open areas of the forest, searching the ground for their varied diet of fruits, roots, shoots, earthworms, lizards, snakes, and rodents. However, with a clear line of sight, the leopard (p.55), one of the mandrill's chief enemies, will make use of the forest's open runways to launch itself in deadly pursuit of its exposed prey.

Another colourful mammal that can be found rooting around forest glades in west and central Africa is the bush pig (see right). These animals are easily identifiable by their red colouring and white facial markings, which serve to intimidate adversaries.

A unique feature of the bush pig's distribution is that it occurs on the African mainland as well as in Madagascar (pp.60–61), making it the only genus of terrestrial mammal that exists in both regions. It is an able swimmer, which has enabled it to populate islands in Lake Victoria, but it is thought to have reached Madagascar by floating on beds of reeds or other buoyant vegetation.

Spiral-horned antelopes are well represented in Africa's forests. Although many species prefer the protection offered by long grasses and heavy undergrowth, all need to drink; so they make their way to forest clearings at dawn and dusk to drink from exposed

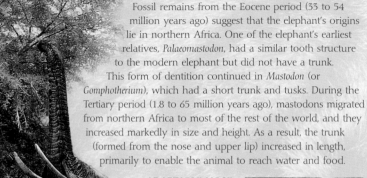

Fossil remains from the Eocene period (33 to 54 million years ago) suggest that the elephant's origins lie in northern Africa. One of the elephant's earliest relatives, *Palaeomastodon*, had a similar tooth structure to the modern elephant but did not have a trunk. This form of dentition continued in *Mastodon* (or *Gomphotherium*), which had a short trunk and tusks. During the Tertiary period (1.8 to 65 million years ago), mastodons migrated from northern Africa to most of the rest of the world, and they increased markedly in size and height. As a result, the trunk (formed from the nose and upper lip) increased in length, primarily to enable the animal to reach water and food.

ADAPTABLE NOSE
The elephant uses its trunk as a tool to reach food and water, as a shower hose, as a snorkel, and as a means of defence.

pools of water. The smallest spiral-horned antelope is the bushbuck, which, like the bongo (p.55), has vertical white stripes on its sides and flanks (although the definition of these markings diminishes towards the eastern and southern limits of the range). Forest pools attract many mammal species, ranging from the eland, the largest of the spiral-horned antelopes, to the tiniest insectivores, such as the elephant shrew.

BUSH OR "TUFTED" PIG
Apart from rooting for bulbs and insect larvae, the bush pig will follow chimpanzees in order to take advantage of the fruit they drop from the trees.

CONFLICTS

Protecting territory or the cohesion of a family group, mating rivalries, and the threat of physical danger, may all lead to disputes between forest dwellers. Cats scent-mark their territory and will fight intruders that ignore these warnings. Like gorillas (p.56), chimpanzees involved in intra-community squabbles will make intimidating gestures and vocalizations, and such disputes are usually settled relatively quickly. However, unlike gorillas, chimpanzees are proactively aggressive towards chimps belonging to outside groups: parties of adult males will scout their territory for intruders and physically confront any that they find. Such encounters may result in severe injuries to individuals on both sides and may even end in death.

Mandrills and bush pigs (see left and above) initially rely on their facial markings to intimidate potential attackers or rivals into withdrawal, but both species are readily prepared to support this with more forceful confrontation if necessary. However, such displays of physical strength or aggression are no match for guns and machetes and, armed with these, man will always be the greatest threat to the wildlife of Africa's forests.

AFRICAN BUFFALO
Oxpeckers pick off and eat the parasites infesting an African buffalo's skin. The buffalo's forest-dwelling subspecies, the African forest buffalo, found only in west Africa, is smaller than its savannah counterpart.

AFRICAN GOLDEN CAT
About twice the size of a domestic cat, this heavily spotted (or wholly black) predator feeds on tree hyraxes, birds, rodents, and even monkeys and small antelopes.

Both predators and prey have a vested interest in self-concealment. A leopard's spotted coat enables it to blend into patches of dappled sunlight on the forest floor as it stalks monkeys or small hoofed mammals such as the bay duiker. Conversely, the sitatunga (p.48), with its long brown coat and striped back, melts effortlessly into the background of the forest swamp. When alarmed, this species will submerge itself in water with only its nostrils above the surface.

BAY DUIKER
Also called the red forest duiker, this small, nocturnal, arch-backed antelope conceals itself in hollow trees or dense thickets during the day.

A distinguishing characteristic of one of the lemurs, the aye-aye, is the morphology of its hands. All of the fingers are long, with curved claws, but the middle digit is noticeably extended and consists of little more than skin, bone, and tendon. The aye-aye uses this manoeuvrable digit not only to clean its face meticulously and to comb itself but to catch food: it does this by tapping on the bark of a tree and listening for the sound of a wood-boring grub; it then breaks open the bark with its teeth and probes the wood to extract the grub.

A SENSITIVE PROBE
The aye-aye's long middle finger is the ideal tool for probing for insects and grubs.

FORESTS OF MADAGASCAR

Lying 400km (250 miles) across the Mozambique Channel off Africa's southeast coast, the island of Madagascar covers an area of 587,041 square km (226,658 square miles). A refuge for many rare mammals, Madagascar contains a variety of forest habitats ranging from the high mountain forests found in the central areas, to the deciduous forests of the western and northern parts, and the rainforests of the east coast. Many tree species are similar not only to those found in nearby Africa but to those in Malaysia, South America, and Australia – a diversity that still presents botanists with a conundrum today.

Owing to its historical separation from Africa (and what would later become India) when the ancient continent of Gondwana broke up, the fauna of Madagascar has been allowed to develop in isolation. Although the island's wildlife bears some similarity to that of Malaysia, many specialized forms that are endemic to the island have evolved. In addition to the many species of lemur found in Madagascar's forested areas, the island is home to a variety of bats, rodents, insectivores, and carnivores. This diversity is exemplified not only by mammals but by the abundant species of fish, butterflies, and reptiles.

LEMUR PARADISE

Lemurs form part of the order of primates and are referred to as "prosimians". Some of their early ancestors developed into apes and monkeys, but lemurs themselves evolved separately into their current forms. This evolution has taken place principally on Madagascar. The island's geographical isolation and its abundance of natural habitat have enabled lemurs to

VERREAUX'S SIFAKA
With arms held out for balance, the distinctive Verreaux's sifaka bounds gracefully across clearings and open ground.

FOSSA
The largest carnivore in Madagascar the Fossa inhabits wooded savannah and forest areas up to 2,600m.

subsist with relatively little interference. One of their few predators is the fossa (see below), a sleek and slender carnivore that hunts by day as well as by night. A powerful predator and agile climber, the fossa is a solitary animal whose taste for domestic livestock as well as for lemurs has led to its persecution, which is partly responsible for its endangered status. Habitat loss also plays a part in its declining numbers.

All species of lemur are endemic to the island, which supports 43 species divided into three families: the mouse and dwarf lemurs; the sportive lemurs; and the "true" lemurs. Together with the family Indriidae (avahis, sifakas, and indri) and the family Daubentoniidae (aye-ayes – see box, left), the lemurs form the infraorder Lemuriformes. The most widely known of the "true" lemurs is the ring-tailed lemur (see opposite and box, below), which inhabits the dry scrub, deciduous forest, and spiny forest of south and southwestern Madagascar. The rocky terrain of the Andringitra Massif in the southeast of the island also supports an isolated population. Another easily identifiable member of the Lemuriformes is Verreaux's Sifaka (see left), whose creamy white pelt and distinctive brown crown and nape of the neck distinguishes it from all other species.

CONSERVATION MEASURES

Madagascar's rich inventory of plants and wildlife is under threat from man. Many of the poorer local people have little alternative but to make use of the forest's natural resources, clearing great tracts of land for construction or for pasture for their cattle. Once the tree cover has been removed, the soils are rapidly washed away by the rains and, as the land becomes eroded and unable to support human needs, it is abandoned in favour of newly cleared forest. Such destruction of habitat leads to the reduction of some of Madagascar's (and the world's) rarest mammals. To combat this trend and to protect the environment, 22 national parks have been created, improved ecofriendly farming methods are being implemented, and sensitive tourism is being encouraged.

The ring-tailed lemur, is easily recognizable by its fox-like face, bright orange-brown eyes, and long, black-and-white ringed tail. Shortly after sunrise, it wakes up in its nocturnal quarters in the branches of a tree shared with others and begins the day's hunt for food. These lemurs are highly sociable and move in groups of at least five or six, although a normal group usually consists of 20 or so individuals. They may cover up to 600m (2,000ft) in a day, pausing every 200m (660ft) or so to feed. Having eaten their fill of leaves, fruits, flowers, figs, and bananas, which are gathered with their dextrous hands, these lemurs (unlike most other species) will sit on their haunches with their arms outstretched and enjoy a period of sunbathing.

CAT LEMUR
The ring-tailed lemur is occasionally called a cat lemur on account of the purring and mewing sounds it produces.

LEMURS

The loud vocalizations of the grey gentle lemur can pierce the thick bamboo forests it inhabits. Weasel lemurs and ruffed lemurs are more at home in the rainforests, whilst the brown lemur is more flexible, inhabiting both dry deciduous and moist montane forest.

GREY GENTLE LEMUR

WEASEL LEMUR

BROWN LEMUR

RUFFED LEMUR

TAIL SIGNALS
The ring-tailed lemur's black-and-white tail is highly distinctive. Males transfer scent to their tails so that they can use them in both visual and scent signalling.

NORTH AMERICA

500 MILLION YEARS AGO North
America lies on the equator,
rotated 90° compared to its present
orientation, with ocean all around,
and much of its land flooded.
Fossils of Cambrian sea creatures
are preserved in rocks called the
Burgess Shales, now located in the
foothills of the Canadian Rockies.

200 MILLION YEARS AGO All
major land masses are joined as
the supercontinent Pangaea. North
America lies in the northwest, butted
to Europe on its northeastern edge,
Africa is to the southeast, and South
America is located along what is
now the Gulf of Mexico.

100 MILLION YEARS AGO Large
areas of the central continent are
covered by shallow seas and
lagoons, resulting in fossil-rich
rocks containing remains of marine
reptiles, such as plesiosaurs, in
regions now far inland.

50 MILLION YEARS AGO North
America, which was joined to
Europe in the east as part of the
northern supercontinent of Laurasia,
is being pushed west as the Atlantic
Ocean begins to widen.

5 MILLION YEARS AGO Intermittent
land bridges begin to form, allowing
interchange of mammal species
between North America, Central
America, and South America.

PREVIOUS PAGE:
GREY WOLF
The grey wolf once roamed through
much of North America, but it is
now restricted mainly to the north.

HABITATS OF
NORTH
AMERICA

STRETCHING FROM THE ARCTIC CIRCLE
TO THE TROPIC OF CANCER, NORTH
AMERICA DISPLAYS A HUGE VARIETY
OF CLIMATES AND HABITATS – SOME STILL
PRISTINE, OTHERS TOTALLY DOMINATED
BY HUMANS.

ARCTIC TO TROPICS

North America is home to a vast,
complex mixture of habitats and to
many of the world's best-known wild
mammals. Much of the history of its
wildlife was shaped by the last great
ice age, which ended 20,000–10,000
years ago. This covered the northern
half of the continent with ice sheets and
glaciers, and pushed the climate zones
southwards nearer the equator. As the
ice retreated, some plants and animals – but by no
means all – moved back to reclaim the north again.

The interior plains of the continent's centre are vast,
rolling grasslands, shading to desert in the southwest,
arid rocky scrub southwards, and almost every kind
of wood and forest to the east. In the north lie great
tracts of pines, spruces, firs, and other conifers of
boreal forest, which historically covered more than
one-third of the continent's land area and spread
southwards along the Rocky Mountains chain. There
are also smaller patches of unique habitats, such as the
temperate rainforest strip along the northern Pacific
shoreline, fed by moist ocean winds, and diagonally
opposite across the continent, the lush, evergreen-
dotted, subtropical swamplands of Florida.

INCREASING POPULATIONS

North America has had intermittent land links
with Asia for many millions of years, via what
is currently the Bering Strait, and with South
America for just the past few million years.
This has allowed numerous mammal species
to enter the continent and spread, including
various incoming waves of humans. Compared
to other continents, North America as a whole is
not packed with people – Europe has more than three
times its population density. But the industrial power,
unfettered enterprise, and technological advances shown
by its people in the past two centuries have vastly
affected its countryside and wildlife, including many
of its most treasured mammals.

POLAR (PP.90–93)
Within the Arctic Circle, snow
and ice whiten the landscape
for more than half the year.
The fiercest conditions are
found on the open, treeless
lands of the tundra, bordering
the Arctic Ocean, parts of the
northern islands, and Hudson
Bay. From here, ice sheets
spread far out across the sea
in mid-winter.

MOUNTAINS AND ARID (PP.72–75)
The Rocky Mountains form an extensive
stony backbone down the west of the
continent, providing a remote refuge for
many larger mammals such as pumas, or
mountain lions. The land descends to dry
scrub and desert, for example the Sonora in
the southwest, then rises again as the bare,
arid Western Sierra Madre range in Mexico.

A R C

Bering Strait

Aleutian Islands

Bering
Sea

Aleutian Range

Mount McKinley
6194m

Alaska Range

Brooks Range

Be

Mackenzie
Delta

Mackenzie Mountains

Gr

PACIFIC

Gulf of Alaska

Coast Mountains

Great

R O C K Y M O

Coast Ranges

Cascade Range

Sierra Nevada

Great
Basin

Great Salt
Lake

Colorado

O C E A N

Mojave
Desert

Grand
Canyon

Sonoran
Desert

Appalachians

Lower California

Gulf of California

Sierra Mad

P A C

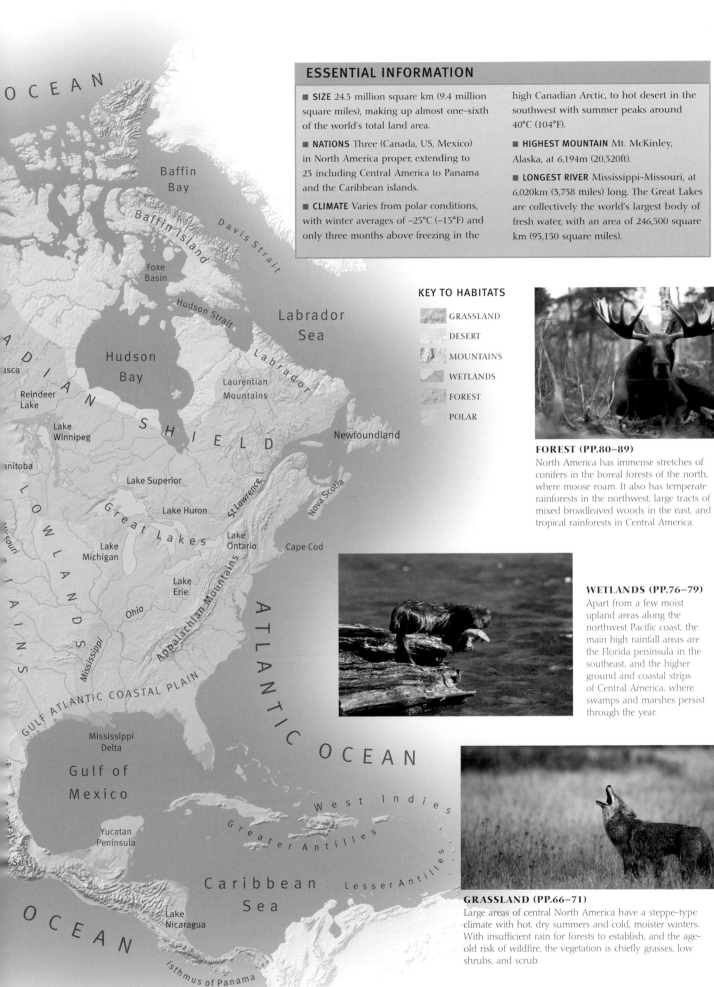

OCEAN

Baffin
Bay

Baffin Island

Davis Strait

Foxe
Basin

Hudson Strait

Labrador
Sea

CANADIAN SHIELD

Hudson
Bay

Laurentian
Mountains

Labrador

asca

Reindeer
Lake

Lake
Winnipeg

Newfoundland

anitoba

LOWLANDS

Lake Superior

Great Lakes

St Lawrence

Nova Scotia

Lake Huron

Lake
Ontario

Cape Cod

Lake
Michigan

Lake
Erie

Ohio

Appalachian Mountains

ssouri

Mississippi

ATLANTIC OCEAN

GULF ATLANTIC COASTAL PLAIN

AINS

Mississippi
Delta

Gulf of
Mexico

West Indies

Greater Antilles

Lesser Antilles

Yucatan
Península

Caribbean
Sea

Su

Lake
Nicaragua

OCEAN

Isthmus of Panama

ESSENTIAL INFORMATION

■ **SIZE** 24.3 million square km (9.4 million square miles), making up almost one-sixth of the world's total land area.

■ **NATIONS** Three (Canada, US, Mexico) in North America proper, extending to 23 including Central America to Panama and the Caribbean islands.

■ **CLIMATE** Varies from polar conditions, with winter averages of –25°C (–13°F) and only three months above freezing in the high Canadian Arctic, to hot desert in the southwest with summer peaks around 40°C (104°F).

■ **HIGHEST MOUNTAIN** Mt. McKinley, Alaska, at 6,194m (20,320ft).

■ **LONGEST RIVER** Mississippi–Missouri, at 6,020km (3,738 miles) long. The Great Lakes are collectively the world's largest body of fresh water, with an area of 246,500 square km (95,150 square miles).

KEY TO HABITATS

- GRASSLAND
- DESERT
- MOUNTAINS
- WETLANDS
- FOREST
- POLAR

FOREST (PP.80–89)
North America has immense stretches of conifers in the boreal forests of the north, where moose roam. It also has temperate rainforests in the northwest, large tracts of mixed broadleaved woods in the east, and tropical rainforests in Central America.

WETLANDS (PP.76–79)
Apart from a few moist upland areas along the northwest Pacific coast, the main high rainfall areas are the Florida peninsula in the southeast, and the higher ground and coastal strips of Central America, where swamps and marshes persist through the year.

GRASSLAND (PP.66–71)
Large areas of central North America have a steppe-type climate with hot, dry summers and cold, moister winters. With insufficient rain for forests to establish, and the age-old risk of wildfire, the vegetation is chiefly grasses, low shrubs, and scrub.

GRASSLAND TYPES
Natural rolling plains in the continental centre turn green with new growth in spring and herbs and multicoloured flowers in early summer. To the west, the upland pastures of the Rocky Mountain foothills are covered in snow through winter, while the southwestern chaparral is dense, tangled brushwood.

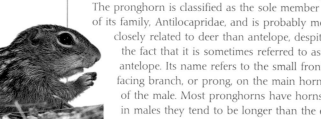

GRASSLAND THE WIDE-OPEN PRAIRIES
OF CENTRAL NORTH AMERICA ARE A SEEMINGLY ROMANTIC HABITAT, BUT THE REALITY IS MORE FORBIDDING – HOT AND DROUGHT-RIDDEN IN MIDSUMMER, SCOURED BY ICY WINDS AND SNOWSTORMS IN WINTER, AND BLEAKLY WINDSWEPT AT ALMOST ALL TIMES.

LIFE IN THE OPEN
Africa's savannas (pp.30–39) survived recent ice ages relatively unscathed and, with year-round warmth, boast dozens of different antelopes, gazelles, and other bovids, as well as equids (horses) such as zebras. However, similar grassy habitats in North America were greatly reduced by the ice ages and also endure much colder winters. These are two reasons why the prairies have relatively few species of large grazing mammals. The biggest are the bison (pp.68–69), along with a few kinds of open-country deer and, on a smaller scale, the pronghorn (see above and p.68).

PRONGHORN OR "PRAIRIE GHOST"
The pronghorn has legendary powers of speed and stamina, easily outpacing any carnivore. It feeds on various grasses, herbs, and even cacti.

The pronghorn is classified as the sole member of its family, Antilocapridae, and is probably more closely related to deer than antelope, despite the fact that it is sometimes referred to as an antelope. Its name refers to the small front-facing branch, or prong, on the main horn of the male. Most pronghorns have horns: in males they tend to be longer than the ears, and in females they are shorter than the ears, or of equal length. Males also have blackish faces and neck patches, which females lack. The horns themselves are also unusual, being composed of a permanent bony core, as in antelopes, covered by a bony sheath that is shed and regrown yearly, as in deer.

TAKING FLIGHT
Pronghorn stand about 90cm (36in) tall at the shoulder. They are often regarded as among the fastest of all land animals, able to sprint at over 80km (50 miles) per hour and race along at not much less for many consecutive minutes. One of the pronghorn's major predators out on the prairie is the highly adaptable coyote (see opposite), a smaller cousin of the grey wolf (pp.62–63, 90). The

13-LINED GROUND SQUIRREL
This is a bold and active inhabitant of the central plains bordering Canada and the US.

WARNING COLORATION
BEHAVIOUR

About six species of skunk frequent plains, meadows, forest, woodland, and similar habitats across North America. All skunk are boldly patterned in black and white – some with spots, others with stripes or larger white areas on the back. This striking effect is a warning to predators, since the skunk's speciality in self-defence is to spray foul-smelling, skin-irritating fluid from the anal glands below its tail, if possible into the adversary's eyes. The odour is so strong, the victim may have trouble breathing. Before expelling the fluid, the skunk stamps its front paws, raises its tail, and walks quickly to and fro with a curious stiff-legged gait – all as a warning that spraying is imminent.

READY TO FIRE
This striped skunk arches its tail over its head as a sign that it may suddenly release the foul-smelling, oily fluid, which it can spray up to 2m (6½ft).

coyote has long been regarded as solitary, with a mournful howl to attract mates in the wilderness, but some are now known to form established breeding pairs. They may even gather as a loose pack, like wolves, although with a less strict social hierarchy.

The coyote is a true opportunistic omnivore and will consume anything from grass and fungi to meat, for example rodents such as rats (p.71) and gophers (p.70), as well as the highly threatened prairie dogs (see box, right) and other species of ground squirrel (see opposite).

> "The coyote has long been regarded as solitary, with a mournful howl to attract mates in the wilderness, but some are now known to form established breeding pairs."

VICTIMS OF FARMING · HUMAN IMPACT

Probably North America's rarest smaller mammal, the black-footed ferret has an extremely restricted diet – prairie dogs. Wholesale decimation of prairie dog populations, especially by poisoning, to reduce its destruction of farmland, made the ferret extinct in the wild in the 1970s. This was partly owing to the great reduction of prairie dogs as food, but also because the ferrets consumed poisoned victims, and were consequently harmed themselves by the poisons that built up in their bodies in a process called bio-amplification. Outbreaks of diseases, including plague in the prairie dogs and canine distemper in the ferrets, also wreaked havoc on populations. Captive breeding programmes and recent re-introductions of the ferrets in Wyoming, South Dakota, and Montana will hopefully re-establish the species in the wild.

PREDATOR AND PREY
The fortunes of the black-footed ferret (left) depend on conserving its main prey, prairie dogs (right).

HOWL OF THE WILD
The howls of coyotes echo across the plains at dusk. The calls allow individuals to keep in touch, proclaim their territories, and attract mates for breeding.

HOW THE MIGHTY FELL

Before European settlers reached the prairies, American bison numbers were estimated at 50 million, perhaps more. They were, undoubtedly, the dominant large grazers of the interior plains. However, by the start of the 20th century they were almost all gone, slaughtered for meat, hides, and other body parts, to make way for livestock. After tremendous conservation efforts, the numbers of the plains bison subspecies probably approach 200,000 today, although more than 150,000 of these are in private collections in parks or commercially reared on ranches. The wood bison subspecies to the north of the continent was always far less common than the plains bison, and is currently regarded as threatened by the Canadian authorities,

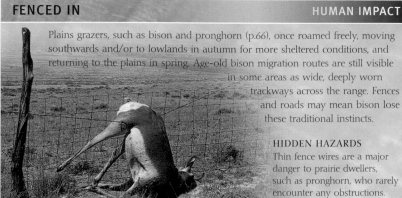

FENCED IN
HUMAN IMPACT

Plains grazers, such as bison and pronghorn (p.66), once roamed freely, moving southwards and/or to lowlands in autumn for more sheltered conditions, and returning to the plains in spring. Age-old bison migration routes are still visible in some areas as wide, deeply worn trackways across the range. Fences and roads may mean bison lose these traditional instincts.

HIDDEN HAZARDS
Thin fence wires are a major danger to prairie dwellers, such as pronghorn, who rarely encounter any obstructions.

with numbers of about 3,000. Also, recent genetic studies show that the European bison, or wisent (p.184) may be the same species as the American bison.

APPEARANCE AND BEHAVIOUR

The bison is North America's heaviest land animal, and one of the biggest of all bovids. A well-grown male may stand over 2m (6½ft) at the shoulder and weigh more than 1 tonne (2,200lb), with a large female about three-quarters this size. Both sexes have short, dark, curved horns, with a shaggy mane over the head, neck, and shoulders. The mane is slightly more conspicuous in the male, as is the shoulder hump. A typical bison herd consists of cows, their young calves, and probably also juveniles, all under the

dominance of a senior female. A few bulls may follow the herd, while others form their own small bachelor groups. The herds roam slowly about 3km (2 miles) per day to locate fresh grazing.

Ruminating in the peace of the plains or wallowing in mud, bison seem unconcerned. But they are ever-alert, with keen hearing and sense of smell, and they are able to detect danger from several kilometres away. Once roused, they stampede with surprising speed. Some larger adults even leave the herd to confront the enemy, lowering their heads and pawing the ground to convey a charge threat. Natural predators, especially of the calves, include grey wolves (pp.62–63), coyotes (p.67), and pumas (pp.72–73).

RUTTING BISON
Bison bulls clash in late summer, issuing challenges by bellows heard 5km (3 miles) away, and then butting and head-ramming rivals.

BISON STAMPEDE
Despite their shaggy bulk, bison can thunder along in excess of 60km (40 miles) per hour – either away from a predator, or towards it, to trample it into the dust.

GRASSLAND DWELLERS

DEER, such as the mule, or black-tailed deer, adapt to different diets through the seasons. They feed on grasses, shoots, and herbs in summer, and twigs, nuts, fungi, and lichens in winter.

FOXES, such as the swift fox (above) of the central prairie, and the extremely similar kit fox just to the west, are omnivorous and dig dens about 4m (13ft) long and up to 1m (3ft) deep.

DEER MICE thrive in North American grasslands, where at least 20 species occur. Their populations reach great densities in favoured habitats, which include bluestem grasses.

GOPHERS live under-ground, where they dig an extensive burrow system, consisting of 30m (100ft) or more of tunnels, enabling them to access grass roots and bulbs.

DIGGING FOR DINNER
The American badger is a solitary nocturnal hunter, especially of grassland rodents such as rats, voles, gophers, and ground squirrels, which the badger digs out of burrows at great speed.

POPULATION CYCLES BEHAVIOUR

About seven species of cotton rats live across North America, mostly in grassy or shrubby habitats. They eat almost anything, from grasses and seeds to insects and worms, at almost any time, day or night. Some years they reach plague numbers and take to water for fish, frogs, crayfish, and the eggs and young of water birds, especially bobwhite quail. Population numbers rise and fall like this naturally over several years, but the cycles are exaggerated in farming areas, where the rats consume various crops, including sugar cane and sweet potato.

TRAIL OF THE RAT
The hispid cotton rat of southern North America can climb reeds with ease to access birds' nests or food stores. It makes well-worn foraging trails in the grass.

PLAINS DEER

Most deer are woodland browsers, but two species in North America have adapted partly to plains life, although they prefer areas where clumps of trees offer scattered shelter. The two are very similar, and in captivity they may interbreed, but in the wild they rarely do so, owing to subtle differences in habitat and behaviour. The mule or black-tailed deer (see opposite) prefers slightly drier conditions and ranges further west, from the plains to the Rocky Mountains and down into the arid scrub and desert of southwest North America. Its tail is either brown-black or white on the upper side, with a dark tip, and white beneath. The white-tailed deer (p.80) is more adaptable, living in moister grasslands to the east, and in forests and marshy regions. Its tail has a dark central stripe edged by white on the upper side, with white below. There are also antler differences between the two species, with the mule deer's dividing into two nearly equal branches, while the white-tailed deer's branches are noticeably unequal. Unusually for deer, neither species tends to gather in large herds, except perhaps during a hard winter. The usual grouping is a doe with her young, or small gatherings of bachelor males.

SMALLER GRASSLAND DWELLERS

Many other grassland dwellers have preferences for grassland sub-habitats, ranging from the drier thorn grasses to the damper, lusher meadow grasses. The black-tailed jackrabbit (see right) inhabits the former and can cope with the heat of the southwest, while the meadow vole (see right) prefers the equable east, and is often found near water. Like most small grassland rodents, these voles burrow for shelter and are prolific breeders. The female has her first litter at less than one month old, and thereafter may have up to ten litters annually, with up to ten young in each. However, pressures from predators, such as the American badger and foxes (see opposite), as well as owls, hawks, and cats, are so great that the chances of a female surviving a full year at such breeding capacity are very small.

BLACK-TAILED JACKRABBIT
This species is born fully furred, with eyes open. It is soon ready to flee its nest – a shallow depression in the ground.

MEADOW VOLES
Meadow voles are born helpless in a grassy nest in the undergrowth or in a shallow burrow.

ANCIENT HOME OF THE HORSE EVOLUTION

The evolution of horses, or equids, is one of the best-known and most-studied areas of palaeontology. The tendency through time has been to larger body size but fewer toes per foot. Each foot originally had five toes, each with a small, nail-like hoof on the front. The first and fifth toes shrank away first, then the second and fourth, to leave just the middle toe completely capped by the large hoof. North America has been home to many kinds of horses, from the rabbit-sized, forest-browsing *Hyracotherium* of 50 million years ago, to the elegant, pony-sized *Hipparion* of the open grasslands, only two million years ago. But the great glaciations of recent ice ages wiped out all horses from North America. Evolution continued in central Asia (see box, p.157), where the modern species was domesticated a few thousand years ago. Horses only returned to North America with European newcomers during the 1500s.

MESOHIPPUS
This prehistoric horse stood about 60cm (24in) high at the shoulder, and lived in North America about 30–25 million years ago. It had three toes on each foot.

GONE WILD
Mustangs are the feral, or "gone wild" descendants of domesticated horses, brought to North America by European settlers *c*.1520.

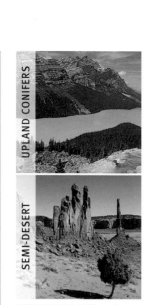

MOUNTAIN AND ARID TYPES
The northern Rocky Mountains support evergreen forests on higher slopes. But lack of rain to the south means scrubby semi-desert shades to patches of true desert.

MOUNTAINS AND ARID
THERE ARE MANY UPLANDS IN NORTH AMERICA, INCLUDING THE ROCKY MOUNTAINS. AS THEIR NAME SUGGESTS, THE ROCKIES ARE MOSTLY BARE OF VEGETATION, PARTICULARLY IN THE DESERT-LIKE SOUTHERN REGION, WHERE RAINFALL IS ESPECIALLY LOW.

WESTERN SPINE
The Rocky Mountains form the western backbone of North America, stretching for almost 5,000km (3,000 miles) north to south, with east–west dimensions varying from 130–650km (80–400 miles), and elevations ranging from 1,500m (4,900ft) to the peak of Elbert, Colorado at 4,399m (14,432ft). To the Rockies' north are the Alaskan and Mackenzie Ranges, with the Coastal and Cascade Ranges to their south flanking the Pacific shores, the Sierra Nevada in the southwest, and the Sierra Madre farther south still. These topographies are mostly rugged and, in the southern half of the continent, highly arid. As a result, mammal life is limited – but often big and spectacular.

As with mountains of other continents, the Rockies have become a remote refuge for larger predators driven from the lowlands by persecution,

MOUNTAIN LION TERRITORY
In the mountains, where prey is scarce, one puma may roam an area greater than 1,000 square km (400 square miles). In general, male areas overlap female territories only.

especially by livestock-owners. Lynx, bobcats (pp.84–85, 86), pumas (see opposite and right), coyotes (p.67), brown and black bears (pp.74–75, 81), grey wolves (pp.62–63), and the large mustelid known as the wolverine (p.81) all still dwell here, as well as in the great boreal forests to the northeast (p.80).

The puma is one of North America's most enduring survivors. Originally widespread in many habitats, it had the greatest range of any mammal in both North and South America, although this has contracted greatly over the past century. Its various names include cougar, panther, and of course mountain lion. With a head-body length of up to 2m (6½ft), and weight of 100kg (220lb), this is the largest of the "non-big" cats. Indeed, it exceeds some so-called big cats, such as the clouded leopard (p.169), and has about the same bulk as the leopard. The puma has

a proportionately small head, powerful legs, and very large paws for its body size. It leaps easily among the crags – it can jump 5m (16ft) vertically onto a ledge or branch. Staple prey are small mammals such as rats, ground squirrels (see right), hares, rabbits, and pikas (see right). But it may also hunt larger prey including bighorn (p.74), Rocky Mountain goats, deer, and occasionally farm animals. In some regions, mule deer, or black-tailed deer (p.70) form well over half of all their food.

The puma is a mostly solitary animal, although males and females pair for the mating season, generally in late autumn, and may stay together for longer. It is relatively vocal for a cat, making an eerie scream when courting, and a variety of hisses, growls, low purring, and curious, flute-like "whistles".

COPING WITH WINTER

Apart from their southernmost regions, the Rocky Mountains are blanketed with snow and ice during winter. In autumn, wild sheep and goats move to the shelter of forests on the lower slopes, followed by their predators, such as grey wolves, coyotes, pumas, lynx, bobcats, and wolverines.

Smaller mammals cannot cover such distances on migration. One option is to stay put but to save energy and reduce exposure to danger by becoming inactive. North American pikas – compact-bodied, shorter-eared relations of rabbits and hares (order Lagomorpha) – do just this. There are 26 pika species, all Asian apart from two in mountainous western North America. Each is based in a den in a jumble of rocks and stones, with surrounding meadows and pastures providing food. In autumn, each pika gathers grasses and herbs into a "haystack", which provides a source of food for the winter. The yellow-bellied marmot (see above), a large ground squirrel of the Central Rockies, stores food inside its body rather than outside. The marmot fattens itself up in autumn, eating as much as it can, to enable it to hibernate for up to 8 months of the year.

YELLOW-BELLIED MARMOT
This bulky rodent weighs up to 5kg (11lb) in autumn, ready for the long winter hibernation. It lives in a small colony, usually one male with several females.

"[The puma] leaps easily among the crags – it can jump 5m (15ft) vertically onto a ledge or branch."

MOUNTAIN DWELLERS

SOME CATS, such as pumas (above), kill their prey and rarely scavenge, while others – lynx and bobcats for example – tend to feast on carcasses.

GROUND SQUIRRELS, for example, the Colombian ground squirrel, eat various plants as well as grubs and worms.

PIKAS, such as the North American pika, vigorously defend their territories and inflict serious wounds on intruders of their own kind by biting and scratching.

ENERGY CONSERVATION   BEHAVIOUR

In arid mountains and desert habitats, two essential resources are scarce – food and water. Many mammals have complex behavioural strategies for coping with these shortages. Sunbathing in the cool season warms the body and so saves burning food for heat energy. In hotter weather, relaxing in the shade reduces the need to cool the body by sweating or panting, both of which cause the animal to lose valuable moisture.

SHADE-BATHING
The desert cottontail of North America's southwest quarter prefers sagebrush scrub with scattered rocks or bushes for shade. It feeds in the cool of the evening and night.

BIGHORN
This sheep grazes quietly in mountain meadows and pastures, but if it senses danger, it leaps away with tremendous speed onto nearby rocky bluffs or cliffs. Its wide, rimmed hooves can grip even wet rocks covered with slippery lichens.

GRASSHOPPER MOUSE
Among the most carnivorous of all rodents, grasshopper mice feed on insects, spiders, scorpions, and other mice. Three species inhabit the dry west and southwest.

SHEEP AND GOATS

Two well-known bovids living at high altitudes in northwest North America are the Rocky Mountain goat and the bighorn (see left). The former inhabits the northernmost Rockies and ranges into Alaska, while the latter is found farther south and east, to the foothills of the Great Plains. The two mammals are close relatives, both belonging to the subfamily Caprinae, or goat-antelopes. Their distributions overlap in the north-central Rockies, and both endure similar conditions, with much the same foods and predators.

However, there are major differences between the two mammals. The Rocky Mountain goat has long, fine, white-yellow fur, while the bighorn's coat is shorter and brownish. The goat's horns are small, sharp, dark, and slightly curved, while those of the bighorn, as its name implies, are massive, light-coloured, and spiralled. The horns of a large male may measure more than 1m (3ft) around their curve and weigh up to 15kg (33lb) – perhaps one-fifth of his total body weight. The female's horns are smaller and less spiralled.

The contrasts extend to behaviour, too. Rocky Mountain goats fight over mates or food only rarely, yet when they do, the battles are serious and may involve both males and females. They jab their

BLACK BEAR
The American black bear turns over stones in streams to find crayfish, crabs, and other freshwater food, which it scoops up with its short-clawed but powerful paws, or sucks up with its mobile, prehensile lips.

BROWN BEAR
Bears roaming a regular route leave signs, such as fur on a tree or rock, claw marks on bark, a stone, or a post, and piles of droppings.

horns into an opponent's flanks and sometimes inflict bloody wounds, although they never intentionally clash heads. Bighorn rams indulge in much more ritualized and lengthier, but in the end less harmful, contests during the rutting season. They face each other, then walk away, before turning around, and lunging forward and down, smashing their heads and horns with tremendous force, the crash resounding through the rocky peaks. This form of battle may continue for a few hours until one of the rams feels too dazed to continue and retreats.

BEARS OF MANY COLOURS
The wild sheep and goats of the Rockies occasionally fall prey to the largest predators of the mountain region – the bears. The biggest is the brown bear, or grizzly bear (see left and p.81), which roams many habitats across all northern continents and is found also in southern Europe (pp.184–85) and Himalayan Asia. The American black bear (see above), at up to 300kg (660lb), is unique to the continent and reaches about half the brown bear's bulk. More than 90 per cent of its diet is plant-based, from roots and tubers to fruits, nuts, and berries. Despite its name, the American black bear is by no means this colour throughout its range. To the west of the Rockies it may be yellow-brown or cinnamon, and along the Pacific coasts, a steely shade of grey-blue.

FOODS FOR ALL SEASONS
Both bear species have an extremely adaptable diet and take advantage of almost every seasonal flush of food. In spring, they consume tender shoots, buds, and stems. In summer, their dietary needs change and a more carnivorous menu is required, with the bears often developing a taste for young sheep, goats, and deer. Autumn is their time of plenty. Not only do they feast on fruits, berries, nuts, and the grubs and honey of bees, but they also trek to upland rivers for the annual "fish run", when they feed on salmon and trout (see opposite). The fish, after swimming and leaping from the sea to their spawning grounds in mountain streams, are exhausted and easily caught. The bears almost queue up to hook them from the water with their long-clawed paws. In early winter, the bears return to their dens for a long sleep.

HABITUATION
BEHAVIOUR

North America's vast natural areas are increasingly frequented by tourists, travellers, and trekkers. Opportunistic mammals, from raccoons (pp.82–83) to bears and even moose (p.89), especially when young, quickly learn to beg snacks from well-meaning but misguided visitors. The animals become "habituated", or accustomed to human presence, which can affect their natural behaviour and survival skills and sometimes leads to trouble if they approach less welcoming people.

BEGGING BEAR CUBS
Bear cubs are endearing as the original "teddy bears". However, they will grow into large, powerful carnivores with much less predictable reactions.

"The fish…are exhausted and easily caught…The bears almost queue up to hook them from the water with their long-clawed paws".

EASY CATCH
Bears soon learn the locations of rapids, pools, and shallows along the rivers in their area, where fish such as salmon and trout are easiest to catch. A big brown bear may eat 50kg (110lb) in a few days.

HUNTING FROM THE RIVERBANK
The mink's eyes are less well adapted for underwater vision than those of the otter. As a result, the mink tends to watch from a rock or bank, and then dive onto passing prey.

WETLANDS
NORTH AMERICA HAS ALMOST EVERY AQUATIC HABITAT, INCLUDING ARCTIC LAKES THAT ARE FROZEN NINE MONTHS OF THE YEAR, TORRENTIAL RAPIDS THAT RUSH DOWN MOUNTAINSIDES, WIDE, LAZY RIVERS, SUBTROPICAL SWAMPS, AND MANGROVE-FRINGED LAGOONS.

SWAMPS AND MARSHES

One of North America's best-known wetlands is the Everglades, originally some 10,000 square km (3,860 square miles) of subtropical swamp and marsh in southern Florida. This unique habitat supports many fascinating mammals, including the fully aquatic, plant-grazing West Indian manatee, or sirenian (pp.110–11) and, on land, one of the rarest subspecies of puma, the Florida panther, with an estimated wild population of just 50-70. The marsh rabbit (see right) is another Florida-dweller, nibbling sedges and reeds, and constructing an above-ground nest of stalks and stems. It, too, has a critically threatened subspecies, Hefner's marsh rabbit of Florida's far south.

LAKES AND RIVERS

Much more extensive than the exotic outpost of the Everglades, are the large rivers and great lakes of the continent's north and east - including the Great Lakes themselves, which contain almost one-fifth of the world's surface fresh water. North America does not have fully aquatic freshwater mammals, like the river dolphins of Asia or South America's Amazon (p.110), but there are many semi-aquatic types. Two of the main groups are the rodents – including the beaver (pp.78–79) and the muskrat (p.78) – and the mustelids, such as otters and mink (see opposite and box, right).

UNWELCOME EXPORTS

Mink and muskrat are two of the continent's less welcome mammal exports. In the 1900s, American mink were established on fur farms in Europe, Asia, and South America, but individuals soon escaped into

MARSH RABBIT
Members of the cottontail group, marsh rabbits have large, splayed toes that enable them to swim fast and run rapidly on soft mud.

local wetland habitats. The mink's introduction caused much disruption to local wildlife on these continents. Not only does it feed voraciously both on land and in water, taking rats, mice, voles, birds and their eggs, amphibians, fish, crayfish, and just about anything edible; it also occupies territories along riverbanks and lakeshores, sometimes at the expense of its rarer relatives, including its extremely threatened, slightly smaller close cousin, the European mink, and some species of otter. The mink's tendency to hunt more on land than the otter has led it to confrontation with poultry farmers and other breeders of small livestock. The muskrat is a large member of the vole subfamily. Like the mink, it was transported to Europe for fur farming, escaped, and has spread widely. It was also deliberately introduced into the wild in northern Asia with a view to being hunted for its fur. It is an expert swimmer, able to cover more than 100m (330ft) under the water or stay submerged for more than 15 minutes. Its varied diet includes not only plant matter, such as reeds and sedges, but also aquatic animals such as fish, frogs, and shellfish. The muskrat burrows prodigiously, and may construct a nest similar to a beaver's lodge (pp.78–79).

"The mink feeds **voraciously** both on **land** and in water, taking rats, mice, **voles**, birds and their **eggs**, amphibians, **fish**, crayfish, and just about **anything** edible."

RIVER RAPIDS

REED-FRINGED LAKE

WETLAND TYPES
Wetland habitats are shaped by the speed of currents, which determine plant growth. In streams foaming rapids make a hazardous environment with little vegetation, while reed-fringed pools and lakes are rich in aquatic insects, amphibians, fish, and the mammals that hunt them.

OVERHUNTING AND CACHING
BEHAVIOUR

Presented with plentiful food, a mink naturally takes advantage by killing more than it needs, with the intention of caching excess victims for later consumption. When prey such as chickens are numerous and confined, the mink faces an unnatural situation. It may continue to kill, apparently wantonly, but it is really following its natural hunting instincts.

MINK'S CACHE
Mink follow the natural behaviour pattern of many predators, killing prey to cache and eat later.

MUSKRAT
The muskrat sometimes makes a lodge-like nest at the water's edge or on a small island. This is smaller and simpler than the beaver's version, consisting chiefly of mud and water plants, and measuring up to 1m (3ft) high and almost 2m (6½ft) across.

THE AMERICAN BEAVER

No mammal of comparable size has changed the natural landscape as much as the American beaver (see right and opposite). Vast tracts of central and northern North America have streams and pools where once there were none, owing to the beaver's industry. Weighing approximately 25kg (55lb), and with a head-body length of 80–90cm (32–36in), plus an additional 30cm (12in) for the wide, flat, scaly-skinned tail, this is the third largest rodent, after the capybara of South America (pp.106–107) and the American beaver's European counterpart, which can weigh over 30kg (66lb).

AQUATIC ADAPTATIONS

The beaver shows many adaptations for life in water which are shared with other semi-aquatic mammals such as otters. The feet are webbed, for powerful swimming. The nostrils and ears close with valve-like flaps when submerged. The fur is long and dense, and kept scrupulously groomed (see opposite). The whiskers are also long and profuse, for feeling the way in cloudy water or at night.

The beaver also has additional adaptations for life under the water which are possessed by few other mammals. It can close its lips behind its large front incisor teeth, to nibble or gnaw like a typical rodent, but under the water. The eyes each have a transparent "third eyelid", the nictitating membrane, which closes to protect the eyeball's surface when diving but still allows the beaver to see. And the tail is almost unique among mammals. One of its main functions is to make a warning slap or splash on the surface of the water when the beaver senses danger and dives beneath. The sound alerts other members of the beaver's family to take evasive action.

On land, the beaver waddles awkwardly on small front legs, with hips raised up on the larger rear legs. The feet are angled inwards under the body, giving a pigeon-toed gait, and its head is held low to the ground.

DAM AND LODGE BUILDING

Beavers usually live in family groups and eat a wide variety of plant food, including buds, shoots, leaves, twigs, and bark of trees, for example aspen, willow, alder, birch, and poplar. They gnaw through trunks wider than 20cm (8in)

BEAVER COUNTRY
Beavers use the lake to escape from land-based predators, and to store food under the water for winter. Before making their lodge they usually make a dam (above) out of twigs, branches, mud, and stones.

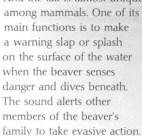

ADAPTING TO WATER ▐ EVOLUTION

Unusually for a mole, the star-nosed mole of eastern North America is adapted to life in water as well as on land. It lives around swamps, marshes, and along the banks of ponds, pools, streams, and rivers. Like other moles, it digs tunnels on land for its nest, for travelling, and for some food-gathering. However, it feeds mainly in water. Its unique nose has 22 fleshy, incredibly sensitive tentacles to feel for small aquatic prey. The mole's front limbs are shovel-like and adapted for digging, its rear feet have long, wide toes for swimming, and its long, furless, scaly-skinned tail is similar to that of the beaver, but narrower.

FEELING WITH FINGERS
The star-nosed mole's finger-like tentacles feel among mud and stones along the bottom of the lake for small animals such as insect larvae, young fish, water snails, and leeches.

FAMILY LIFE
Beavers live in small groups, usually consisting of an adult male and female, and their kits (offspring). There are 3–4 kits per litter, born fully furred and with eyes open.

thick to fell the tree and enable them to reach the tender buds and young leaves. In addition to felling trees for food, beavers utilize the wood for two major construction projects. First, they build a wall–like dam (see opposite) using twigs and branches, which they pile up at a suitable point across a waterway, reinforced with mud and stones. Most dams are about 20m (65ft) long but some extend for more than 500m (1,650ft). Water piles up behind the dam to form a deeper pool, where the beaver can then erect its dome-shaped lodge. In areas where there are natural deeper

pools, beavers rarely bother with dam construction. The lodge is also made from twigs, boughs, pebbles, and mud. Its base can be more than 10m (33ft) across, and its height above the water surface over 2m (6½ft). The lodge's entrance is under the water and leads to a den chamber within, which is above water level and lined with dry vegetation. The lodge is a safe retreat from terrestrial predators, such as wolves and cats. It is also a shelter from severe weather and a breeding den in spring.

DAILY GROOM
The beaver grooms for at least an hour a day to get mud out of its fur and to spread water-proofing oils over its body.

FOREST

ABOUT 500 YEARS AGO, APPROXIMATELY TWO-THIRDS OF NORTH AMERICA WAS CLOAKED IN WOODS AND FORESTS. IN RECENT CENTURIES, MANY AREAS HAVE BEEN LOGGED AND CLEARED. HOWEVER, EXTENSIVE TRACTS REMAIN, AND MANY ANCIENT DECIDUOUS WOODS ARE NOW PROTECTED.

FOREST TYPES

The vast boreal forests of the north include pines, spruces, firs, and larches. Broadleaved woodlands in the east turn brilliant colours in autumn. Swamp cypresses, laden with moss and creepers, thrive year-round in the warm, damp south.

NORTHERN FORESTS

A journey across North America would passs through most of the world's major forest types. In the north is the vast boreal forest, which is coniferous. This massive wilderness is the stronghold of many of North America's most impressive predatory mammals, including pumas (pp.72–73), bears (see opposite and pp.74–75), grey wolves (pp.62–63, 90), the wolverine (see opposite), and medium-sized cats such as lynx and bobcats (p.84–85, 86). It is also home to some of the largest browsers, such as moose (see opposite and pp.88–89) and wapiti (see opposite). Conifer coverage also extends southwards, down the west of the continent, along the rugged slopes of the Rockies. The low temperatures and rot-resistant needle-leaves in boreal forest mean decomposition of the leaf litter into nutrient-rich soil is slow For this reason, small ground mammals, such as voles and shrews, are much less numerous than in the deciduous woodlands to the south.

DECIDUOUS WOODLAND

The continent's most extensive areas of broadleaved woodland are in the east and southeast, from the Great Lakes and across the rolling Appalachians. Here, deciduous trees such as maple, oak, beech, mahogany, hickory, and hornbeam lose their broad leaves in a blaze of autumn yellows, golds, and reds. These woodlands are home to a miscellany of smaller mammals, including several species of squirrel, for example the grey squirrel (pp.82, 189), woodchucks or groundhogs (see opposite), rats, mice, rabbits, opossums (p.86), foxes (p.86), shrews (see opposite) and of course numerous types of bat. Many of these mammals, together with adaptable species such as white-tailed deer (see left) and raccoons (pp.82–83), are also found in the far southeast, as the trees change to magnolias and

WHITE-TAILED DEER

A stag sports his antlers, ready for the autumn rut. These deer tend to be larger in northern forests than southern ones.

"The vast **boreal** forest is the stronghold of many of North America's most **impressive** predators, including **pumas, bears,** and **wolves.**"

NATURAL CLEARANCE

Open patches fringing a forest pool are visited regularly by large herbivores such as moose, who gradually enlarge them to bigger clearings (inset) by their nibbling and browsing.

subtropical evergreens such as pines, cypresses, and palms. A few larger predators, such as black bears, pumas, and bobcats cling on in these areas.

WESTERN RAINFOREST

A more restricted habitat is the narrow strip of temperate moist forest along the Pacific coast of the northwest. This is dominated by the massive redwoods, or sequoias. The region has several specialized mammals, including the Douglas

AMERICAN SHORT-TAILED SHREW
Found mainly in broadleaved eastern woodlands, this is a large member of its group, with a head–body length of up to 15cm (6in), but only 3cm (1¼in) of tail.

squirrel, the American shrew-mole – really a mole but with many shrew-like features – and the mountain beaver. The mountain beaver is regarded as one of the most primitive of all rodents. In sharp contrast to the better-known American beaver (pp.78–79), the mountain beaver lives more like a mole, digging a complex of tunnels 100m (330ft) or more in length. It resembles a large vole, and emerges at dusk to gather various plant foods.

FOREST DWELLERS

BEARS, such as the brown (grizzly) bear, were originally natural woodland dwellers, exploiting numerous plant foods with the seasons.

MUSTELIDS, for example the wolverine, hunt small mammals and birds, scavenge larger carcasses, and eat plants.

GROUND SQUIRRELS, such as the woodchuck (groundhog), hibernate in their burrows in winter.

DEER, such as wapiti, browse trees and graze low grasses and herbs. They range different areas with the seasons.

COMMON RACCOON
The raccoon's appearance is unmistakable: a long, ringed tail, and fox-like face, with dark eyepatches merging to form the "bandit" mask.

RACCOONS

One of North America's most familiar and adaptable mammals is the common raccoon (see below and opposite). It is the best-known representative of its genus, *Procyon*, which includes two other species from the Central and South American mainland and islands. These are all members of the mammal family Procyonidae, which also includes coatis (p.109) and kinkajous (p.120). Raccoons are related to dogs and bears, but their original home is the trees – especially in damp broadleaved woodland with streams and marshes. However, the common raccoon has widened its range greatly during the past century, mainly because of its association with human habitation. It is now found in almost every environment from riverbanks and swamps to scrub and semi-desert, and also even in urban environments such as parks, gardens, and backyards.

SPECIALIST GENERALIST

The common raccoon has been called a "specialist generalist". It specializes in being a general opportunist in almost every way, including anatomy, behaviour, and diet. Its build is stocky and muscular yet agile, and it can climb trees, fences, and even walls

WATCHING OUT

Young raccoons are playful, energetic, and keen to investigate. In natural woodland areas they often make dens in hollow trees. If tree holes are lacking, they may excavate burrows.

BREEDING HABITS   BEHAVIOUR

The female raccoon builds a nest in a sheltered site, perhaps under a building or among rocks, and after a gestation period of 60–70 days, she gives birth, usually in April, to three or four offspring – although there may be up to seven. Young raccoons usually stay with their mother for at least 6 months, and in the northern areas perhaps until next spring. Then they disperse to set up their own home ranges.

NEWBORN RACCOONS
The youngsters' eyes open after about 3 weeks, and by the time they are 2 months old they begin to venture out with their mother on their first foraging trips.

with ease, using its long, slender toes for gripping, and its long tail for balance. It is also a reasonable swimmer and regularly takes to water when hunting or to escape danger. The common raccoon is active at any time, day or night, according to its habitat and food availability. It is bold and inquisitive, and readily investigates into crevices, under logs, and down holes in the search for food. Its diet is extremely wide-ranging. Broadly, its prey falls into three groups: aquatic creatures, such as fish, frogs, water snails, crayfish, and insect larvae; smaller land-based animals, such as mice, voles, and other rodents, as well as birds' eggs and chicks, grubs, snails, and worms; and finally plant matter, especially berries, seeds, fruits, and nuts.

FEELING FOR FOOD

The raccoon's front paws have long, dextrous "fingers", which are used to scrabble and dig, and feel for prey in loose soil and in the sand or mud on the bottom of streams and pools. The raccoon often seems to "wash" its food by swishing or shaking it in water, although this appears to be an in-built or instinctive action, related to grubbing for food in watery mud rather than actually washing food because it is dirty. In fact, the name "raccoon" is said to derive from native North American terms *arakun* or *aroughcan*, which infers that the animal "scratches with its hands".

The raccoon's manual dexterity and adventurous, inquisitive nature allows it to investigate novel sources of food, which often brings it into conflict with people. It can "fiddle" with latches and locks to open them and gain access to chicken coops or similar enclosures, where it happily feeds on eggs (see right). Fresh corn is another raccoon favourite, and one that does not endear the mammal to farmers.

AT HOME IN OUR HOME

The common raccoon's range has extended steadily northwards through the past century. This has coincided with clearing land for farming and human habitation. Raccoons are a familiar nuisance around towns, where they forage in trash cans and garbage tips, making noise and mess. They also establish dens in sheds, barns, and other outbuildings, and may breed here too.

In the wild, the raccoon is generally solitary for most of the year, apart from a short mating season in January and February, and perhaps a communal den for the depths of winter, where several raccoons stay inactive for long periods but do not truly hibernate. However, the concentration of foods and denning sites around farms and towns can bring 20 or more raccoons together into bands or packs, when they can be very destructive.

Unlike many other carnivorous mammals, raccoons do not have well-defined territories which they defend vigorously from others of their kind. But each raccoon does have a home range, which it roams regularly. It seems that the home ranges of males do not overlap, but a male's range may overlap with the ranges of two or three females, and the ranges of females may partly overlap. However, meetings are rare since raccoons tend to avoid each other. In typical woodland, the home range is usually 30–70 hectares (75–170 acres). However, in food-rich areas, especially around human habitation, a range can be just a few hectares, while in dry grassland and scrub, with sparse food availability, a range may exceed 1,000 hectares (2,500 acres).

FLEEING GREY SQUIRREL
Raccoons forage in tree branches and take the young of squirrels, birds and their eggs, and almost any tree-dwellers as prey.

BREAKING EGGS
Raccoons may push or knock eggs from a nest so that they fall to the ground and crack, or throw them at nearby logs or stones. They then open the shells fully with their paws and lick out the contents.

FINDING FOOD
The raccoon uses its front paws, or "hands", to locate food by feeling with a scrabbling motion. Muddy banks of pools and streams are favourite foraging places.

WINTER CHASE
The lynx lunges desperately as the hare leaps to escape. This split second may decide two lives, since the lynx could begin to starve now that snow lies thick on the ground.

"The American or Canadian lynx moves through snow easily on its long, wide-pawed feet, and is a voracious predator, especially of snowshoe hare."

FOREST PREDATORS

The bobcat, named after its short or "bobbed" tail, inhabits central and southern regions and feeds mainly on rabbits. The grey fox is a nocturnal predator in woodlands in the south and consumes a wide variety of small creatures, fruits, and berries. The ringtail, also known as the cacomistle, is found only in the US Mexican plateau. It feeds on small animals, flowers, fruits, sap, and nectar. The fisher lives in a broad east–west zone, mainly across southern Canada and the northern US. It feeds on mammals and birds.

BOBCAT **GREY FOX** **RINGTAIL** **FISHER**

FOXES AND CARNIVORES

The medium-sized predators of North American forests (see box, above) represent several subgroups within the major mammal group, Carnivora. From the family Canidae there are several kinds of foxes, including the very widespread red fox (pp.131, 188) and the more woodland-based grey fox (see above). The grey fox climbs expertly and often lives in tree holes, perhaps more than 10m (30ft) above the ground. There are also many representatives from the almost exclusively American raccoon family, including the common raccoon itself (pp.82–83), the kinkajou (p. 120), and the ringtail or cacomistle (see above). The ringtail has raccoon-like eye patches and a ringed tail but is smaller and slimmer, resembling a cat, and it is extremely fast and agile in the branches.

MANY MUSTELIDS

More than 20 species of generally long-bodied, short-legged, short-tailed mustelids roam North America's woods and forests. Some are expert climbers, while others hunt mainly on the ground. They include the weasel (p.91) and stoat (p.179), which are found in all northern continents, and also various skunks, such as the eastern spotted skunk (p.89), otters, the American mink (p.76–77), the wolverine (p.81), and the American marten. The American marten has a head–body length of 40cm (16in) and longer, stronger legs and a bushier tail compared to stoat-shaped mustelids. These features indicate that it spends more of its time in branches than down burrows. Its major prey includes squirrels, such as the grey squirrel (pp.83, 189) and northern flying squirrels (see opposite).

The marten's close cousin, the fisher (see box, above), is more typically mustelid-shaped but considerably larger than the marten, with a head–body length of 60–70cm (24–28in), plus a tail half as long again. Its name is misleading in that it does not prey largely on fish or even in water. Instead, it hunts mainly on the ground, catching various mammals and birds. It is one of the few predators to tackle porcupines (p.89), again on the ground rather than in branches. Its technique is to leap and dart around the porcupine, trying to keep face to face as the porcupine attempts to wheel around and use its spiny quills, while also dashing in to bite the prickly rodent's nose, lips, and eyes. Gradually, the wounded porcupine weakens and the fisher can rush in to rip at its softer underside. It is estimated that a fisher gets over 2 weeks' nourishment from one porcupine, while a squirrel will last a day.

CHIPMUNK
These bold ground squirrels are staple prey of carnivores in North America. They live alone in burrows but often feed in groups.

FOREST CATS

Two medium-sized, short-tailed cats stalk North American forests. The American or Canadian lynx has a northerly range through the boreal forests, moves through snow easily on its relatively long, wide-pawed feet, and is a voracious predator, especially of snowshoe hares (p.90). The bobcat (see box, above) lives in the centre and south of the continent, has shorter ear tufts and legs, a darker-spotted coat, and its main food is cottontail rabbits.

PLAYING DEAD

The Virginia opossum is the largest marsupial in all of the Americas, and is the only marsupial in North America. It is one of the few marsupials whose range has spread in the past century. Its curious method of self-defence is to feign death, sometimes known as "playing possum". The creature falls down and remains completely still, without moving or responding to any stimulus, with its eyes and mouth open. Many predators that would instinctively tear up a struggling prey become confused, lose interest in the animal, and wander away.

DEAD STILL
A Virginia opossum "playing dead" lies partly curled on its side and does not respond to anything.

NIGHT FLIGHT
Unusually for squirrels, the
northern flying squirrel feeds
mainly at night. It has huge
eyes that enable it to see in
the dark. Its gliding abilities
evolved to enable it to escape
from predators in the trees.

AUTUMN RUT
Rutting moose prepare to push each other in their display of strength. Their antler shape is known as palmate, with up to 20 small points extending from a central broad part, the beam.

THE MIGHTY MOOSE

The largest herbivores of North American forests are deer, but people visiting from Europe, where the same species are found, are sometimes puzzled as to their correct names. The largest of all deer, *Alces alces*, is known as the moose in North America and the elk in Europe and Asia. The name elk is used in North America for another sizeable species, *Cervus elaphus*, which is also called the wapiti (p.81); confusingly, this species is known as the red deer in Europe. (Some experts still hold the view

WOODLAND CLEARING
Moose consume most kinds of plants, including grasses, reeds, and sedges in clearer patches among the trees.

that the American elk is actually not the same species as the Eurasian red deer and should have a different scientific name, *Cervus canadensis*.)

Whatever its nomenclature, the male moose is a huge and impressive beast, reaching 3.5m (11ft) from nose to rump. It is over 2m (6½ft) tall at the shoulder and weighs more than 750kg (1,650lb). His huge antlers are grown anew yearly, as in all deer, and may span 2m (6½ft) from tip to tip. The antlers are fully developed for the rut in autumn (see above), when the massive males bellow, toss heads, and engage in pushing contests. These tend to happen in isolated twos and threes, since the moose is a solitary animal rather than a herd dweller.

The female moose may reach only half of the male's bulk and, as in nearly all deer species, she usually lacks antlers. However, females do have the elongated muzzle and top lip, and the furred fleshy flap hanging from the throat, known as the bell or dewlap, which give moose their distinctive appearance. Unusually among deer, females are active during the rut, too. They display and moan to attract males, and may even have their own clashes.

Moose are fast and powerful, able to run in excess of 50km (30 miles) per hour, swim fast rivers with ease, and kill a wolf with one lashing kick of the broad, sharp-edged hoof. Their favoured habitat is a mix of deciduous trees, especially poplar, birch, and willow, as well as some conifers, and marshy areas, pools, and slow streams where aquatic plants thrive. This deer may wade nostril-deep into the water to chomp on water lilies and similar succulent vegetation, often submerging its head to reach the roots and runners on the bottom.

SPIKY WOODLAND DWELLERS

Another distinctive mammal of northern woodlands is the North American porcupine (see right), although its range extends much farther south than that of the moose. Its formidable method of self-defence consists of thick, sharp, modified hairs, known as quills, which are up to 8cm (3in) long in the crest on the head.

Porcupines are members of the rodent group. They climb well but slowly, and eat a wide variety of plant food, including buds and blossoms in spring, stems and leaves in summer, nuts and berries in autumn, bark and conifer needles in winter, and farm crops seemingly at any opportunity. Porcupines become extremely vocal as they court in early winter, making a variety of pig-type grunts, loud screeches, owl-like hoots, and moaning whines.

DANGERS ON THE ROAD

As more people visit woods, forests, and other wilderness regions, hazards increase for the wild mammal occupants. Creatures in these areas learn that cars often mean food, and as a result they wait around on roadways, lodges, and picnic sites, attempting to gain food by increasingly persistent begging, putting themselves as well as people in danger. The animals can become attracted to cars, sometimes even wandering into fast traffic. More than 100 people each year die in the US as a result of road accidents caused by collision with deer alone.

BAD HABITS
A pair of moose push their vast heads through the open window of a car in the hope of being given some food. Vehicles and these great deer can be a dangerous mix on the roads.

PORCUPINE
If threatened, this spiky rodent backs into its attacker, piercing the animal's flesh with its barbed quills. The quills then detach at their base, enabling the porcupine to escape.

STRIPED MARKINGS
The eastern spotted skunk, which is actually striped, has very variable markings, although all possess the white forehead and tail-tip. Like other skunks, it can spray foul fluid in self-defence.

POLAR TYPES
Across most of northernmost North America, great conifer forests continue almost to the Arctic Circle. Beyond are several hundred kilometres of a relatively flat, treeless landscape, tundra, stretching to the shores of the Arctic Ocean. In the northwest are the icy peaks, foothills, and glaciers of the Brooks and Mackenzie ranges.

POLAR

THE WIDE BELT OF TUNDRA FLANKING THE ARCTIC OCEAN IS PERHAPS THE HARSHEST OF ALL TERRESTRIAL HABITATS. TEMPERATURES PLUNGE WELL BELOW –30°C (–22°F) FOR WEEKS, THERE ARE NO TREES FOR SHELTER, DAYLIGHT IS SCARCE, AND FOOD IS OFTEN BURIED BELOW LAYERS OF SNOW AND ICE.

SURVIVING THE ARCTIC
The northernmost parts of North America, including the major islands such as Baffin Island and the great expanse of Greenland, are mostly within the Arctic Circle. Beyond such latitudes the sun does not rise for at least a few days in winter, and temperatures become desperately cold – average daily readings in January and February are –30°C (–22°F) or below. Despite the harsh conditions, mammals find a way to survive here.

Many North American polar dwellers are found in similar habitats in northern Asia, and have some of the most specialized of all mammalian adaptations – not only to the bone-chilling cold, but also to lack of food. Blizzards and white-outs can prevent the search for

ARCTIC GREY WOLF
This is the largest and thickest-furred of this variable species. All-white wolves are more common in the Arctic than elsewhere.

food for several days at a time, and even a drink requires energy, because body warmth is used to melt the ice.

LIFE ON THE TUNDRA
Where the trees of the vast boreal forest fade away, at about the latitude of the Arctic Circle, the landscape opens out to the habitat known as tundra. It is low and wide, and unrelentingly monotonous to the untutored eye – especially in winter, when it is blanketed by snow. The spring thaw reveals bogs and pools, low hummocks of mosses, grasses, and sedges, and the occasional knee-high clump of dwarf willow or birch trees. This is also the time when long trekking columns of caribou (p.92) emerge from the shelter of the forests, on their northward migration

"The spring **thaw** reveals bogs and pools, low hummocks of mosses, **grasses, and sedges,** and the occasional knee-high clump of dwarf willow or birch **trees.**"

COATS OF DIFFERENT COLOURS

Many truly polar animals, including the Arctic fox (see right), and the Arctic hare (see right), as well as the stoats of the far north, change their coats with the seasons. In autumn, the brown or grey summer hairs moult, and are replaced with a thicker growth of white fur. This provides excellent camouflage – as well as greater insulation – once the snow and ice begin to arrive.

SNOWSHOE HARE
This broad-footed hare is found mainly in northern forests. In summer, it matches the carpet of dead needle-leaves between the trees. In autumn, its coat turns white (inset).

to take advantage of the tundra's brief summer bloom. This species – known as reindeer in Europe, where they are mostly domesticated – is the only deer in which both males and females have antlers. The caribou are harried by their chief predators, grey wolves (see opposite), who follow the herds to pick off any straggling old, young, or sick.

Already out on the tundra are musk oxen (see opposite and below). They endure the open land year-round, grazing sedges and grasses in valleys during the summer, and moving to higher ground in winter, where strong winds keep the sparse vegetation of mosses and lichens relatively snow-free. By the start of the 20th century, musk oxen had been hunted almost to extinction by humans for their meat, hides, and horns. However, their numbers are now recovering. Reintroductions across their former range include Alaska in North America and Norway in northern Europe. Although musk oxen resemble wild cattle, they are in fact in a different group of bovids, known as "goat-antelopes" (subfamily Caprinae), and are more closely related to mountain goats and bighorn sheep (p.74). The hairs of their outer or guard coat are among the longest of all mammal fur, exceeding 60cm (24in), and form an all-round, fringe-like "skirt" that shrugs off rain and snow. The furry undercoat is very dense and waterproof. The broad hooves minimize sinking in snow. Both sexes have the heavy horns, which are a symbol of maturity for breeding, for intimidation during herd disputes, and are used as weapons of self-defence against wolves and polar bears (pp.92–93).

SEASONAL FEEDING
Musk oxen feed hungrily on the flush of summer plants across the tundra, and are then able to endure many days without food during winter.

POLAR DWELLERS

ARCTIC FOXES are among the few mammals to survive the tundra year-round, often scavenging along coasts for whale and seal carcasses.

ARCTIC HARES live in the tundra year round, and may forsake their usual herbivorous habits to scavenge carrion if plants are scarce.

MUSTELIDS, such as weasels, tend to keep to forested areas, where their main prey includes lemmings, snow voles, and small birds.

SOME RODENTS, such as the brown lemming, stay active all year, digging tunnels under the snow from their dens to feeding areas.

THE KING OF THE ARCTIC

The polar bear is the foremost predator of the Arctic. It is also just about the world's largest land-based hunter, with a head-body length exceeding 3m (10ft) and a maximum weight of more than 700kg (1,550lb). The brown (grizzly) bear (pp.74–75, 81) can match these statistics but has a much more omnivorous diet, while the polar bear consumes more than 90 per cent meat.

For most polar bears, over two-thirds of their diet consists of seals, in particular the ringed seal but also the bearded seal, the hooded seal, the harp seal, and the walrus (pp.195, 206–207). They will also occasionally feed on fish and small whales. Some polar bears that stray further from the coast or live near human habitation have a more varied diet, and will feed mainly on musk oxen (pp.90–91), caribou (see below), hares, voles, lemmings (p.91), and birds and their eggs. In desperate periods, when seals are scarce, or if the sea freezes far from shore, polar bears may amble tens of kilometres inland, where they will subsist on berries, mosses, and lichens.

MUTUAL HELP
Polar bears and Arctic foxes usually tolerate each other's presence. The sharp-eyed fox may help to locate potential prey, and then gnaw on the leftovers after the bear has fed.

The polar bear has a massive body and a small, streamlined head. It is a powerful swimmer, paddling with its broad paws in the icy water for hours if necessary (see opposite). In a sprint on land, the polar bear can easily overtake a human and even outrun a caribou. The wide feet function as snow shoes to prevent sinking, and their furred soles give non-slip grip on the ice.

SOLITARY LIFE

The polar bear, like most members of the bear family, is mainly solitary. Individuals may gather at rich food sources, such as a stranded great whale, or more recently, garbage tips on the edges of Arctic settlements. Male and female bears liaise briefly in early summer for a couple of days to mate, then go their separate

MIGRATING CARIBOU
As caribou migrate to and from the tundra, swimming across rivers close to the sea, they may be at risk from polar bear attack. Some bears make a habit of moving inland for summer, where they gain a more varied diet.

MALE RIVALS
Sparring polar bears are usually young to middle-aged males fighting over a receptive female. After about 20 years of age, mating activity reduces in both sexes.

SWIMMING LESSON
Cubs hitch a ride with their mother from one hunting ground to another, covering hundreds of kilometres per year. Polar bears paddle at about 10km (6 miles) an hour.

DOZY BEARS
When food is plentiful, polar bears spend up to 90 per cent of their time resting or asleep. Females are able to breed by 5 years, but males, who grow to twice the female's average weight, generally do not tend to mate until they are nine or ten.

ways. The male locates the female by scent – polar bears have extremely keen noses and are said to sniff the odour of a whale or seal carcass in the clear Arctic air from more than 10km (6 miles) away.

Males may become aggressive towards other males in the proximity during the mating season, and they may rear up and slap or box each other with their great forepaws (see left). If a starving male chances upon a female with cubs, he may even kill the young to supplement his diet.

SNOWY NURSERY

Most polar bears shelter occasionally from severe conditions for a few days in a cave, hole, or purpose-dug den of some kind. But in about November, a pregnant female establishes a breeding den, which may be under a rocky overhang or dug into snow drifted against a bank. The den consists of a tunnel 2–3m (6½–10ft) long, leading to a chamber just big enough for her body. Once inside, she is unlikely to leave for four months. The litter size averages two and, as with other bears, the newborn cubs are tiny compared to their mother. They weigh only 600-700g (20–24oz), and curled up they are about the size of small grapefruits, their eyes tightly closed. However they already have a short, velvety version of the adult's white coat.

The mother lives off her body fat as she suckles the cubs for about three or four months, sheltered from the worst of the winter. The family emerges in spring, when the cubs weigh 10kg (22lb) or more. They begin to share their mother's kills and practise hunting for themselves after about 6 months. But they are not weaned for another 6-12 months, and they may not leave their mother for an independent life until their second summer.

HUNTING TECHNIQUES   BEHAVIOUR

The polar bear has various hunting techniques. On land, the bear inches slowly towards its victim, "freezing" if it looks up. It then charges the last few metres in a couple of seconds. At sea, the bear lies very still on the ice near a seal's or whale's breathing hole. When the victim emerges, the bear lunges and bites it hard. The polar bear often puts its paw over its nose and closes its eyes so that its dark spots are less obvious in the white surroundings.

VARIED VICTIMS
A polar bear feasts on its kill (left), and a beluga whale (right) shows the wounds inflicted by a polar bear at the whale's regular breathing hole.

SOUTH AMERICA

HABITATS OF
SOUTH AMERICA

SOUTH AMERICA IS A CONTINENT OF GREAT CONTRASTS,
FROM HYPER-ARID DESERT TO THE LARGEST RIVER SYSTEM
IN THE TROPICS, AND FROM FLAT, DRY PLAINS
TO ONE OF THE LONGEST AND HIGHEST
MOUNTAIN CHAINS IN THE WORLD. IT ALSO
CONTAINS VAST EXPANSES OF TROPICAL FOREST.

130 MILLION YEARS AGO South America is joined to Africa, forming part of the giant southern continent of Gondwana. The southern Atlantic Ocean has yet to form.

65 MILLION YEARS AGO By the end of the Age of Reptiles, the southern Atlantic opens up. Gondwana splits apart, and South America becomes an island, positioned significantly further south than it is now. Cut off from other continents, its mammals start to evolve into distinctly South American forms.

20 MILLION YEARS AGO South America is drifting northwards, and is approaching its current position. The Andean mountain chain already exists, but an episode of mountain building creates even higher peaks, particularly in the northern part of the chain.

3 MILLION YEARS AGO A fall in sea levels, accompanied by continuing uplift of the Andes, creates a land-bridge between North and South America. In a great biological interchange, mammals move between the newly joined continents.

18,000 YEARS AGO Towards the end of the last ice age, the climate is dry, and the Amazon rainforest is considerably smaller than it is now. Low sea levels expose a large shelf of land off Patagonia, and ice caps cover the southern Andes. The remnants of those ice caps still exist today. Humans probably arrive in South America at about this time, but the exact date is not certain.

PREVIOUS PAGE:
NIGHT MONKEY
South American night monkeys, or douroucoulis, are the only fully nocturnal monkeys in the world.

THE TROPICS

At its southernmost point, South America is only 900km (560 miles) from Antarctica, but because of its top-heavy shape, four-fifths of the continent lies in the tropics. The vast Amazon Basin forms the heart of South America – a region of immense rivers, stifling heat, and daily downpours of thundery rain. These conditions are ideal for plant growth, and they nurture the largest rainforest on Earth, albeit one that is shrinking fast.

The richness of Amazonian wildlife is legendary: nearly 90 per cent of all the primates in the Americas live in its vast expanse, and so do the majority of South America's marsupials. In the northeast and southeast of the continent, the Amazon is flanked by the gently rising highlands of Venezuela and Brazil, but to the west, the low ground ends at the Andes. Stretching from the Caribbean coast to the rocky islands off Cape Horn, this spectacular mountain range is home to some of South America's most remarkable mammals, including vicuñas (p.104) and guanacos (p.104), both members of the camel family, as well as animals with adaptations that enable them to breathe easily in the thin air and with thick coats that protect them against the night cold. Rain can be sparse on the lower slopes, and in Peru and Chile, the Andes flank the Atacama Desert – the driest region in the world.

THE DEEP SOUTH

South of the tropics is the region known as the Southern Cone. South America changes character here, as lush forests are replaced first by grassland, and then by the arid and windswept scrub of Patagonia. The largest mammals in this area are introduced cattle and sheep, but native predators such as foxes (p.104) still thrive in the thinly populated landscape. In the far south, the Andes come to an end in a spectacular landscape of jagged peaks, ice caps, and fjords. Here, the foothills are covered with temperate forest – the largest expanse of this kind of vegetation in the southern hemisphere.

Gulf of Darien

Gulf of Panama

Galapagos Islands

Gulf of Guayaquil

Punta Negra

Cordillera Real
Cordillera Occidental

Marañón

PACIFIC

MOUNTAINS (PP.104–05)
The Andes is by far the largest mountain chain in South America. Many mammals live at over 5,000m (16,400ft) in the tropical part. In the far south permanent ice caps cover most ground over 3,000m (9,850ft).

WETLANDS (PP.106–11)
South America's freshwater wetlands include permanent lakes and swamps, and also ones that are seasonally inundated, such as flooded forest or *igapó*. Most of its freshwater mammals divide their time between land and water; dolphins and manatees are the only ones that remain in water all the time.

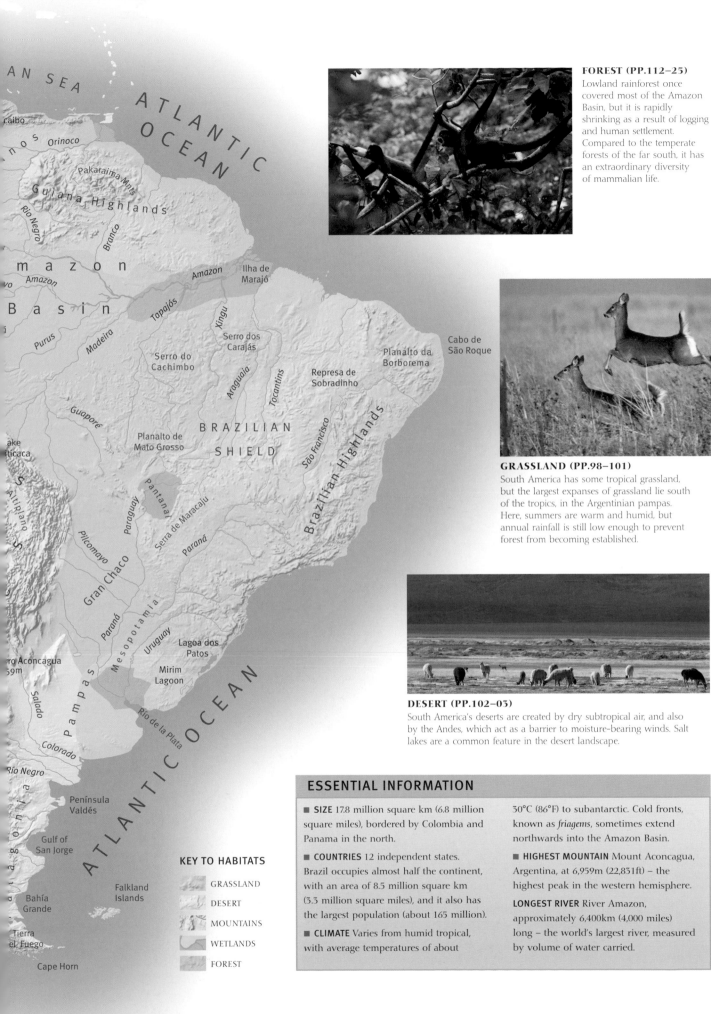

FOREST (PP.112–25)
Lowland rainforest once covered most of the Amazon Basin, but it is rapidly shrinking as a result of logging and human settlement. Compared to the temperate forests of the far south, it has an extraordinary diversity of mammalian life.

GRASSLAND (PP.98–101)
South America has some tropical grassland, but the largest expanses of grassland lie south of the tropics, in the Argentinian pampas. Here, summers are warm and humid, but annual rainfall is still low enough to prevent forest from becoming established.

DESERT (PP.102–03)
South America's deserts are created by dry subtropical air, and also by the Andes, which act as a barrier to moisture-bearing winds. Salt lakes are a common feature in the desert landscape.

KEY TO HABITATS

- GRASSLAND
- DESERT
- MOUNTAINS
- WETLANDS
- FOREST

ESSENTIAL INFORMATION

■ **SIZE** 17.8 million square km (6.8 million square miles), bordered by Colombia and Panama in the north.

■ **COUNTRIES** 12 independent states. Brazil occupies almost half the continent, with an area of 8.5 million square km (3.3 million square miles), and it also has the largest population (about 165 million).

■ **CLIMATE** Varies from humid tropical, with average temperatures of about 30°C (86°F) to subantarctic. Cold fronts, known as *friagems*, sometimes extend northwards into the Amazon Basin.

■ **HIGHEST MOUNTAIN** Mount Aconcagua, Argentina, at 6,959m (22,831ft) – the highest peak in the western hemisphere.

■ **LONGEST RIVER** River Amazon, approximately 6,400km (4,000 miles) long – the world's largest river, measured by volume of water carried.

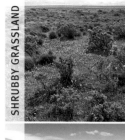

GRASSLAND TYPES
South America's grasslands become drier from east to west, and from north to south of the country. In the drier regions, such as Patagonia, drought-resistant shrubs are commonly found, while in other regions the plant cover is made up primarily of grass.

GRASSLAND SOUTH AMERICA HAS VARIOUS TYPES OF GRASSLAND, WHICH BRING TOGETHER MAMMALS WITH VERY DIFFERENT EVOLUTIONARY ORIGINS. IN THESE AREAS, TYPICAL GRAZERS AND CARNIVORES LIVE ALONGSIDE SOME OF THE MOST UNUSUAL MAMMALS IN THE WORLD.

GRASSLAND GRAZERS

Unlike Africa and Asia, South America has never had any antelopes, although it did have some huge grazers in the distant past. Today, the largest wild grazers are deer. The tropical grasslands of Venezuela and Brazil are home to the white-tailed deer (see below and p.80), a pan-American mammal that ranges as far as Canada. It feeds on grass and shrubs, and usually stays close to cover. Its distant ancestors, like those of all South America's deer, migrated from North America when the two continents joined about 3 million years ago.

Southeast of the continent lies the pampas, the largest expanse of grassland in South America. It covers an area slightly larger than Texas and is a region of open horizons and almost table-flat terrain. Today, much of the pampas is used for growing crops, but in outlying areas many original grassland mammals survive.

The pampas is home to the pampas deer (see opposite). Deer are normally animals of woodland and forest, but this species has successfully adapted to life in the open. It once existed in huge numbers and was as important for the native peoples of the pampas as

ADAPTABLE DEER
Sprinting for safety, a pair of white-tailed deer show off their distinctive white flashes. This species copes with a remarkable variety of climates over its immense range.

the bison was for native North Americans. Since the advent of farming, the pampas deer population has declined sharply, although there are significant numbers in national parks.

Another pampas grazer is the plains viscacha, a large rodent that weighs as much as an American badger. It lives in extensive burrow systems called viscacheras, which are sometimes over 100m (330ft) across.

A single system may be home to up to 50 animals. They emerge to feed at dusk and dawn. Viscachas are famous for their speed and agility above ground and for their tireless digging – something that has brought them into conflict with farmers. They will also collect all kinds of discarded objects and take them underground.

GRASSLAND PREDATORS

South America's mammalian carnivores also arrived during the great interchange between north and south. In grassland, their living descendants include the maned wolf (see box, below), the crab-eating fox (p.109), and the pampas cat (see left). Despite its name, the rarely seen maned wolf is more closely related to foxes than wolves. The crab-eating fox is more common, but its name is almost as misleading. Although it feeds on crabs near coasts and rivers, it is an opportunist, eating whatever it can find. Like South America's other foxes, or zorros, it often scavenges food from dead remains. The pampas cat, on the other hand, is exclusively a hunter of live prey, attacking rodents and ground-dwelling birds, including poultry.

PAMPAS CAT
This small cat lives in such habitats as forest, grassland, swamps, and mountains. It hunts on the ground for prey.

PAMPAS DEER
This deer lives in small groups of up to six animals. The males have scent glands on their back feet, which produce a pungent smell used both to attract mates and to mark territory.

"The pampas covers an area slightly larger than Texas, and is a region of open horizons and table-flat terrain."

SOLITARY HABITS
BEHAVIOUR

Unlike its better-known namesake, the maned wolf is solitary and difficult to observe. Although a male and female often share the same range, they hunt alone. Maned wolves feed primarily on small rodents and birds, and their hunting technique is fox-like, pouncing on their prey rather than pursuing it in a long chase. As with true foxes, their diet also includes some plant food. Maned wolves get their name from their dark-coloured mane, which stands on end if they feel threatened.

LEGGY PREDATOR
Exceptionally long legs enable this pampas predator to see over tall grass.

"The giant claws make formidable weapons if the anteater is threatened by cougars or jaguars."

ANT SPECIALISTS

In grassland, ants and termites can be extremely numerous, but only a few mammals successfully specialize in this kind of food. South America is home to one of the largest small insect specialists: the giant anteater (see below and opposite). It weighs up to 40kg, (88lb) and, with its bushy tail, can be over 2m (6½ft) long. Three smaller species of anteater live in South America's forests – the silky anteater (p.125) being the smallest – and collect their food in trees.

Anteaters have evolved some remarkable adaptations for their unusual way of life. All of them have long, toothless snouts, and ribbon-like tongues that are coated with sticky saliva and covered with backward-pointing spines. When an anteater feeds, its jaws stay almost still, but its tongue flicks in and out at high speed. The tongue sweeps up hundreds of insects a minute, and the anteater swallows continuously,

drinking a mixture of saliva and suspended food. Anteaters are careful about which ants and termites they eat, avoiding ones that are particularly aggressive when attacked. To obtain their prey, anteaters use their front claws to rip open insect nests. The giant anteater's claws measure over 10cm (4in) long, and because it lives on the ground, it has to take special measures to keep them sharp. It walks on the knuckles of its front feet, with its claws turned upwards and out of the way. These claws make formidable weapons if the anteater is threatened by cougars (pp.72–73) or jaguars (pp.122–25) – among the few animals that include giant anteaters among their prey.

TERMITE KINGDOMS

During the last ice age, South America's grasslands were home to armoured mammals called glyptodonts, which weighed nearly 250kg (550lb). The last glyptodonts died out about 10,000 years ago, but their close relatives – the armadillos (see right and p.103) – still survive. There are about 20 species, and many of them feature termites in their diet.

In forests, termites often build their nests from wood fibres, which makes it relatively easy for insect eaters to pull them apart. Climbing mammals, such as silky anteaters (p.125) and tamanduas (p.125), specialize in this kind of food. In grassland, however, the usual construction material for nests is mud, which becomes extremely hard once it has been baked by the sun. Small armadillos have difficulty breaking in, and they usually collect termites that are foraging away from their nests. But the giant armadillo has much stronger claws and legs, and it can smash a nest apart. However, this armoured insect eater does not simply break up nests and move on. Often, it carves permanent burrows beneath termite mounds, creating a refuge that doubles up as a source of food. Giant armadillos feed after dark, and cover about 2.5km (1½ miles) a night.

LONG-NOSED ARMADILLO
Probing among fallen leaves, an armadillo searches for food. As a family, armadillos are found in a wide range of habitats, including grassland, forest, and desert (p.103).

GIANT APPETITE
A giant anteater can eat over 30,000 ants or termites a day, using its sticky tongue (opposite). Females give birth to a single young, which clings to its mother with its claws (above).

BLOOD-FEEDING BATS   BEHAVIOUR

Few mammals inspire such fear in humans as vampire bats. South America is home to three species, which live in a variety of habitats including grassland and forest. All feed on the blood of roosting birds and sleeping mammals, and they approach their prey on all fours, by scuttling spider-like across the ground, or along the branches of trees. Vampires take only small amounts of blood when they feed but they can be dangerous as they sometimes transmit rabies to cattle, and occasionally to humans.

COMMON VAMPIRE BAT
Vampires feed by slicing out a small piece of skin, and lapping up blood as it flows out. An anticoagulant in their saliva stops the blood clotting.

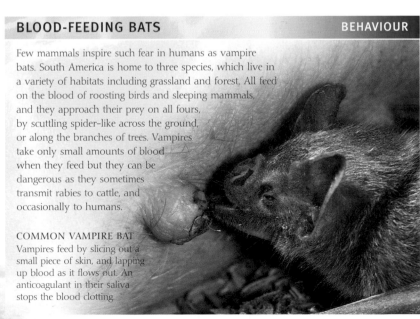

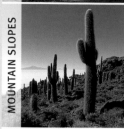

DRY HILLS

DESERT TYPES
South America's deserts vary widely, according to their latitude and rainfall. The central Atacama is hyper-arid, with almost lunar landscapes, while further south, the eastern slopes of the Andes are studded with cacti and scrub. The Andes create a rain shadow that gives the highland area of Patagonia its dry climate.

DESERT
THE ATACAMA DESERT, ON THE COAST OF CHILE AND PERU, IS THE DRIEST PLACE ON EARTH. FEW ANIMALS SURVIVE IN THESE EXTREME, ARID CONDITIONS. HOWEVER, SOME OF SOUTH AMERICA'S DESERTS ARE RICH IN MAMMALIAN LIFE, PARTICULARLY IN THE LESS PARCHED REGION OF PATAGONIA.

SOUTHERN DESERTS

South America has two main desert regions, separated by the mountainous spine of the Andes. The Atacama Desert lies to the west of the mountains, and is wedged between them and the sea. In the central part of the Atacama, rainfall can be as low as 0.1mm (0.004in) a year, which means that there is little plant cover of any kind. To the east of the Andes, and much further south, are the deserts and drylands of Patagonia. Here, where moisture is more dependable, much of the land is covered by drought-resistant scrub.

On the Atacama coast, the only large mammals are ones that feed at sea. Southern fur seals regularly haul out on barren beaches, and marine otters, which are land-based but spend a lot of time in water, hide away in caves that open near the water line. The desert itself contains relatively few mammals, apart from small rodents, such as the degu (p.104), but llamas (see opposite) are a common sight on higher ground. Although they behave with all the wariness of wild animals, llamas are domesticated livestock and have been raised in the Andean region for over 5,000 years. There are two main breeds – the light-fleeced ccara and the heavy-fleeced chaku. Both breeds are raised for their meat and wool, but ccara llamas are still used as pack

COLLARED PECCARY
An animal of forest, savanna, and semi-desert scrub, the collared peccary has a wide distribution, from southwestern US to Argentina. In South America, it is hunted for its meat.

animals, much as they were in Inca times. Llamas often share their arid pasture with wild asses or burros – a living legacy of the Spanish conquest. The natural habitat of the wild ass is Europe's stony semi-desert, so it is hardly surprising that they have adapted well to their new home. In the scanty grassland of the Andean foothills, it is not unusual to see herds of 30 or more, which run off quickly if approached.

On the fringes of the desert, where rainfall is sufficiently high for shrubs to grow, collared peccaries (see left) make up part of the mammal fauna. These pig-like animals eat a wide range of food, browsing on plants with their sharp teeth, or ploughing up the soil with their snouts. Long hunted by humans, the collared peccary also faces a formidable natural predator: the cougar, also known as the puma or mountain lion (pp.72–73). The cougar is a remarkably adaptable animal, equally at home in desert and in the much more closed surroundings of tropical forest.

SAFETY THROUGH SPEED
BEHAVIOUR

The mara, or Patagonian cavy, is one of the few rodents that lives in the open, and that runs for safety instead of seeking shelter underground. Standing up to 50cm (20in) at the shoulder, it weighs up to 16kg (35lb), and has long legs that end in hoof-like claws. It has keen eyesight, and when threatened can run at 45km (28 miles) per hour, hurtling through the scrub with a peculiar bouncing gait. When resting, it either sits on its haunches or lies down and tucks in its legs like a cat. Young maras are born above ground, but they spend their first few weeks in a burrow.

MOTHER AND YOUNG
Maras usually give birth to two young each time they breed. The young are well developed at birth, and may suckle milk from more than one female.

DESERTS OF PATAGONIA

With its dry climate and punishing summer winds, Patagonia is one of the most thinly populated parts of South America. Here, domestic sheep outnumber people many times over, but wild mammals are also common. They include two species of armadillo – the nine-banded armadillo (see box, right) and the long-nosed armadillo (pp.100–101) – and the mara, or Patagonian cavy (see box, opposite), a highly distinctive rodent that looks like a cross between a deer and a hare.

Two species of foxes live in this part of the continent – the culpeo (p.104) and the grey zorro. Both feed on Patagonia's native rodents, but they also help to control two introduced mammals: the European rabbit (p.179) and the brown hare (pp.174–75, 178–79). Since these species arrived in Patagonia in the early 1900s, they have become extremely successful in their adopted home and have become pests in some areas.

BODY ARMOUR
EVOLUTION

Several groups of mammals have body armour consisting of spines or scales, but armadillos are the only ones that are protected by bony plates covered with skin. An armadillo's head, shoulders, and hindquarters are covered by plates arranged in continuous shields, while the plates across its back are arranged in bands separated by flexible skin. When threatened, some armadillos can roll up into a ball, but the majority either pull in their legs or dig rapidly to bury themselves underground. An armadillo's armour protects the animal from attack and also enables it to run through thorny scrub at speed.

NINE-BANDED ARMADILLO
The nine-banded armadillo is widespread and has either eight or nine bands across its back in different parts of its range.

"Although they **behave** with all the **wariness** of wild animals, llamas are domesticated **livestock** and they have been raised in the Andean region for over **5,000 years.**"

GRAZING LLAMAS
On the barren shores of a salt lake, llamas forage for food. Like their relative the camel (pp.40–41, 156), they are good at surviving in extreme conditions.

MOUNTAIN DWELLERS

CAMELIDS, such as the vicuña, are found wild in South America. Despite its dainty appearance, the vicuña is hardy, feeding at extremely high altitudes.

BEARS are frequently more herbivorous than carnivorous. The spectacled bear, South America's only bear, is almost vegetarian.

FOXES are common in the Andes. The culpeo fox ranges along most of their length, except for the far north. It makes dens among fallen rocks.

RODENTS inhabit all altitudes of the Andes. The rat-sized degu is a good digger. It can be a serious pest, because it raids crops and stores food.

ON THE LOOKOUT
Alerted by movement, a guanaco gets ready to make its escape. Long legs and small feet make this animal a good climber in the mountains, and a tireless runner in open country.

MOUNTAINS EXTENDING DOWN THE
ENTIRE LENGTH OF SOUTH AMERICA, THE ANDES CONTAIN THE
HIGHEST PEAKS IN THE WESTERN HEMISPHERE. ANDEAN MAMMALS
ARE ADAPTED TO SURVIVE IN A HABITAT THAT HAS THIN AIR,
INTENSE SUNSHINE, AND SOMETIMES BITTER NIGHT-TIME COLD.

THE CORDILLERA

The climate of the Andes differs greatly from that of
the surrounding lowlands. Even in the tropical regions,
temperatures can drop below freezing after dark, and
rainfall is often low. Forest covers some of the foothills,
but in the main range, known as the Cordillera, there
is little cover of any kind. This abrupt transition has
some interesting effects on South America's mammals.
Primates, for example, extend into the Andean foothills,
but stop where forest ends.

Grassland is uncommon in the steamy heat of
the tropical lowlands, but in many parts of the Andes
it stretches as far as the eye can see. Here, the
largest native grazing mammals are the
vicuña (see opposite) and the guanaco
(see opposite) – two graceful members
of the camel family that specialize in
life high up in the mountains. The
vicuña lives in open country as high
as 5,750m (18,850ft) and is restricted
to the tropics, but the guanaco is found
as far south as Tierra del Fuego, often on
lower ground. Both of these animals have
dense fur and blood that contains more
red cells than most other mammals,
enabling them to take greater levels
of oxygen into their bloodstream. This adaptation
allows them to run with ease at altitudes that leave
humans gasping for breath.

THE ALTIPLANO

In the high plateau of the central Andes, known as
the Altiplano, vicuñas and guanacos share their habitat
with two domesticated species – the llama (p.103) and
the alpaca. All four animals have much in common,
and their close relationship is demonstrated by the
fact that they can interbreed. However, the ancestry
of llamas and alpacas is still a matter of debate. At
one time, both were thought to be descended from
the guanaco, but molecular evidence suggests that
alpacas actually descended from the vicuña. The
vicuña has extremely fine wool and this feature
is even more apparent in the alpaca, which is
raised for its exceptionally luxuriant fleece.

Thick fur is a characteristic of many smaller
mammals that share this high-altitude habitat. Most
famous of all is the chinchilla (see box, right), a rabbit–
sized rodent that lives among rocks at altitudes of up
to 5,000m (16,400ft). Its extremely soft and dense fur
is highly valued by humans, so much so that hunting
and trapping has brought it to the edge of extinction.

A related rodent, the chinchilla rat, also has superb
insulation, although its coat is not quite as woolly.
It feeds on tough vegetation, and it has an unusually
long digestive tract to deal with this kind of food.

The Altiplano was also the original habitat of the
guinea pig (see below). Domesticated at least 3,000
years ago, this adaptable rodent was originally raised
for food – a role it still plays today in many Andean
villages and farms. Many different breeds now exist,
but the original guinea pig has long ago disappeared
from the wild.

GUINEA PIG

Unlike many rodents, guinea
pigs have small families.
They have a long gestation
period, and their young are
well developed when they
are born.

MOUNTAIN CARNIVORES

Cougars, also known as pumas
or mountain lions (pp.72–73), are
widespread throughout the Andes,
but these mountains also shelter
much smaller and rarer cats that
feed on rodents. The smallest is
the kodkod, or guina, which lives
in southern Chile. An animal of forested
foothills and scrub, it weighs as little
as 2kg (4½lb). The Andean mountain
cat is about twice its size and lives in
open, rocky terrain, 5,000m (16,400ft)
high. South American foxes (see opposite)
also hunt mountain rodents, but the largest member
of the Andean Carnivora (carnivores) pays little
attention to animal prey. This is the spectacled or
Andean bear (see opposite) – an inhabitant of forested
foothills and mountain grassland from Venezuela to
Bolivia. It feeds on fruit and bamboo, and animal
food makes up less than 5 per cent of its diet.

GLACIAL VALLEY

HIGH PEAKS

MOORLAND

MOUNTAIN TYPES

South America has a wide
range of mountain landscapes.
In the Andes, these include
steep-sided glacial valleys,
rocky peaks, and moorland
at higher latitudes. The
broadest point of the Andean
chain, in Bolivia and Chile,
embraces a wide plateau
known as the Altiplano.
It is approximately 3,500m
(11,500ft) high.

FUR TRADE   HUMAN IMPACT

Synonymous with warmth and luxury, chinchilla
fur is the most expensive in the world. It was
prized before Europeans arrived in South
America, but after colonization the fur trade
grew rapidly. By 1900, Chile alone exported
500,000 chinchilla pelts a year. Faced
with this huge annual toll, chinchillas
became extremely rare. Today, wild
chinchillas are legally protected, but
hunting continues, and only about
10,000 are thought to remain in the wild.
However, the species is not in imminent
danger, because as many as a million
are raised in captivity for their fur.

DOUBLE COAT

The chinchilla's coat gets its
warmth from the wool that
forms its underfur.

WETLAND TYPES
Most of South America's wetland habitats consist of rivers, lakes, and swamps situated on low-lying and slow-draining ground. However, spectacular waterfalls occur in the eastern highlands of Venezuela and Brazil, while in the Andes, glacial valleys are often filled with deep lakes.

WETLANDS

SOUTH AMERICA'S RIVERS CARRY NEARLY A QUARTER OF ALL THE FLOWING FRESH WATER IN THE WORLD. AS IT SLOWLY MAKES ITS WAY TOWARDS THE SEA, THIS WATER CREATES AN UNRIVALLED VARIETY OF WETLAND HABITATS, FROM BLACKWATER CREEKS TO FERTILE SWAMPS.

THE AMAZON

South America is outstandingly rich in habitats for wetland wildlife because of its size and shape. In the tropics, moist air sweeps westwards across the Amazon Basin, often producing over 2m (6½ft) of rainfall a year. But in the far west, the Amazonian lowlands come to an abrupt end at the Andes, so fresh water has to flow eastwards, crossing an entire continent to reach the sea. The result is the Amazon – a vast river system, full of meanders, backwaters, and lakes, that is larger than any other in the tropical world. The Amazon's level rises and falls throughout the year, but these seasonal changes become even more marked further south,

on the edge of the tropics. Here, the wet season causes extensive flooding between November and March, when the Paraguay River bursts its banks as it flows through southern Brazil and landlocked Paraguay. These floods spill out over an immense and fertile floodplain known as the Pantanal, the single largest wetland in the world.

AT THE FOREST EDGE

As water flows eastwards from the Andes, mountain rivers quickly reach lower ground. From here onwards, the journey slows, and wetland habitats become more diverse. In most places, riverbanks are crowded with

es, but occasionally the forest is interrupted by lakes
d grassy, waterlogged ground. Waterside grass is a
me habitat for the capybara (see below and opposite)
he world's largest rodent, and an animal that invites
mparisons with the hippopotamus (pp.50–52).
hough capybaras are much smaller than hippos,
th a maximum weight of about 65kg (145lb), they
ve a similar lifestyle, using water as a refuge but
ding on land. Capybaras feed mainly on grasses
at grow near the water's edge, cutting them close
the ground with their incisor teeth. They graze
efly in the late afternoon and evening and spend
remainder of the day resting on land, or lounging
the shallows. Capybaras are highly social, and live in
ups consisting of a dominant male and his entourage
females and young. Despite their size, they are vulnerable
attack by predators, including big cats and birds of prey.
the first sign of danger, one animal gives a warning bark,
d the entire group is immediately alert. If the threat
oves real, it rushes for the water – a sanctuary where
predators can follow.

This kind of habitat is also home to the marsh deer
right) – the largest species of deer in South America.
has wide-spreading hooves, connected by a membrane

MARSH DEER
Tall wetland plants conceal
a female marsh deer while
it feeds. Males have antlers
with a double fork. Unlike
deer in cooler climates,
they shed their antlers
at any time of year.

that helps to spread its weight in soft mud. Marsh deer
feed on grass and water plants, and they sometimes
wade chest-deep to reach their food.

IN THE HEADWATERS

South America's most unusual wetland mammal
lives in lowland rivers and also in mountain
streams. Called the water opossum, or yapok,
it is the only marsupial that has adapted to
aquatic life, with its streamlined body and long
tail, each up to 40cm (16in) long, dense, water-
repellent fur, and large, strongly webbed hindfeet.
The water opossum hunts at night and feeds on fish
and crustaceans, feeling for them with its whiskers
and its front paws.

For placental mammals, this diving lifestyle presents
few problems, because breeding females can leave their
young behind in the safety of a nest. But the water
opossum cannot do this, because its young develop
inside its pouch. To prevent them drowning, the
pouch's entrance has a ring of muscle that can
tighten up, sealing off the pouch when the mother
dives. The young can tolerate low oxygen levels,
enabling them to survive once the pouch has closed.

SAFETY IN WATER
A capybara leads her young to join the rest
of the group in the safety of the water. With
slightly webbed feet, they are well adapted to
swimming and can dive for up to 5 minutes.

COYPU
Found in the marshes south of the continent, this aquatic rodent digs burrows up to 15m (50ft) long. In some places, it is bred for its fur.

PACA
The two paca species are long-legged rodents of forested areas near water. Like the capybara (pp.106–107) and the coypu (see above), they are good swimmers, and take to water if threatened.

IN QUIET WATERS

In the countless streams and creeks that feed the Amazon river, the largest mammalian predator is the giant otter (see below) – an animal that can measure over 1.8m (6ft) from nose to tail. This is not only the world's biggest river otter, but it is also the most aquatic. Its paddle-shaped webbed feet give it an awkward gait on land, but in water it is exceptionally lithe and manoeuvrable, a result of its streamlined body, flexible spine, and muscular tail. River otters feed mainly on fish, using their mouths to catch their prey. They breed during the driest part of the year, when river levels are relatively low, and fish become concentrated in the shallows. The young are born in a riverbank den, and they remain with their parents for six months or more. In typical otter fashion, the young are highly active and playful, and the parents often join in, communicating with a repertoire of sounds ranging from quiet purrs to high-pitched yelps. Most river otters do some of their hunting at night – a habit that

GIANT OTTER
This sleek fish eater is about twice the size of most river otters. The anaconda shares its habitat, but the otter's quick reactions ensure the snake is only a minor threat.

is reinforced if humans disturb their daytime hunting. However, the giant otter is entirely diurnal, and its noisy and inquistive behaviour make it relatively simple to track down. As a result, it is an easy target for hunters, and it is now endangered in parts of its range.

FISHING CATS

Domestic cats are notoriously shy of water, as are many of their relatives in the wild. But in the humid American tropics, water is so widespread that no hunter can afford to avoid it. South America's largest cat – the jaguar (pp.122–25) – lives close to lakes and rivers, and far from fearing water, it swims readily and well. It catches fish and caimans, and river otters (see opposite) also feature among its prey. The ocelot (see right) and jaguarundi (p.125) are less keen to take the plunge, but both are proficient swimmers. They also hunt from riverbanks, and during the dry season, they are among the many predators that feast on catfish trapped in shrinking swamps.

OCELOT
Weighing up to 15kg (33lb), the ocelot is one of the largest of the so-called small cats.

The jaguarundi is the least typical South American cat, with a long body, short legs, and a plain or speckled coat. Its overall shape is reminiscent of a weasel, although with its tail included, it can be over 1.2m (4ft) long. The ocelot, on the other hand, has typical cat proportions and an exceptionally handsome spotted coat. Unfortunately, it has paid a price for its good looks: unlike the jaguarundi, it has been a major victim of the trade in furriers' skins (see box, below).

WATERSIDE OPPORTUNISTS

In lowland forest and swamps, other terrestrial carnivores are attracted by the food that can be found at or near the water's edge. They include at least two of South America's foxes or zorros, such as the crab-eating fox (see right), and also the bush dog (p.124) – a short-legged animal that is one of the least known

members of the dog family. Bush dogs often live near lakes and swamps, and they hunt in packs, sometimes swimming well out of their depth to chase their prey.

Throughout the Americas, raccoons are common inhabitants of wetlands, particularly where trees provide cover and places to breed. The common raccoon (pp.82–83) reaches as far south as Panama, but from here the crab-eating raccoon (see right) ranges throughout the Amazon region and beyond. Compared to its northern cousin, the South American species is longer and more slender – an impression emphasized by its much shorter fur. Crab-eating raccoons forage after dark at the water's edge, using their sensitive fingers to feel beneath the surface for food. It used to be thought that raccoons rinsed their food, a myth that resulted from their unusual feeding behaviour.

FRESHWATER BATS

As dusk falls, South America's waterways attract an impressive number and variety of bats. Many of them come to drink, and to feed on insects, but some are in search of larger prey. The fringe-lipped bat, for example, flutters along ditches and streams, where it locates frogs by listening for their calls. It can distinguish different species by the sounds that they make, and snatches up edible ones, leaving poisonous species alone. Bulldog bats (see right), also known as fisherman bats, hunt over larger expanses of calm water, where they use echolocation to pin-point jumping fish. Having located a shoal, the bat then strikes, raking its hind claws through the water before the fish have a chance to swim away. Once it has made a catch, it uses its tail membrane as a safety net as it transfers the fish forwards to its mouth. Fish-eating bats often have well-developed legs and claws, and their swoop-and-grab fishing technique is remarkably similar to the one used by fish eagles and other birds of prey.

WETLAND DWELLERS

FOXES live in a variety of habitats. The crab-eating fox's range stretches from Colombia to Argentina. It often lives in wetlands, but has no special adaptations for life near water.

RACCOONS are relatively common in North America, but the crab-eating raccoon is the only species in South America. Its diet is wide ranging and includes fish, aquatic insects, and plants.

COATIS (here a white-nosed coati) are related to raccoons. They are primarily woodland animals, although they often forage close to water, and sleep in trees at night.

FISH-EATING BATS, such as the greater bulldog bat, have a voracious appetite. Captive bats have consumed 40 fish in the course of a single night.

THE SKIN TRADE
HUMAN IMPACT

During the 1960s, up to 200,000 ocelot skins were exported annually, until the species was included in CITES Appendix 1, which effectively banned the trade. Illegal hunting in ocelots continues, but demand has fallen through changes in fashion and a growing awareness of the ecological effects of buying furs. Currently, up to 3 million ocelots are thought to exist in the wild.

SKINS FOR SALE
Ocelot skins command high prices – partly because of their inherent beauty, and also because no two are exactly alike.

AMAZONIAN DOLPHINS
The bôto's colour changes with age, from bluish grey to pink. Compared to marine dolphins, it has an unusually mobile neck, enabling it to twist its head to examine its surroundings using sound.

DARK WATERS

As water flows eastwards, tributaries merge to form river channels that can be many kilometres wide. In the eastern highlands of Brazil, river water is clear of sediment, although it is stained almost black by tannins washed in from decomposing plants. But in the main branch of the Amazon, the water is highly turbid, making it difficult for animals to navigate or find food by sight.

For the bôto, or Amazon river dolphin, this turbidity is not a problem. At 2.5m (8ft) long, this is the largest of the world's freshwater dolphins. It is also one of the Amazon's largest aquatic predators, matched in size only by the pirarucu, a giant freshwater fish. Like the

four other species of river dolphin (Indus, Ganges, Franciscana, and Chinese), it is practically blind, but its highly developed echolocation system more than makes up for this deficiency. Using sound, bôtos scan the water in front of them and the river bed, and their discrimination is so good that they can neatly remove fish that are trapped in fishermen's nets. When rivers flood bôtos swim into the surrounding forest, steering without difficulty through a landscape thick with semi-submerged trees.

The bôto is a strictly freshwater animal, reaching up to 3,000km (1,860 miles) inland. But another South American dolphin – the tucuxi – is regularly seen in the Amazon and Orinoco estuaries, as well as inland and out to sea. The tucuxi is one of the smallest dolphins in the world, often less than 1.5m (5ft) long.

"Manatees have an **unhurried** lifestyle, **swimming** with slow **beats** of their **paddle-shaped** tails."

GENTLE BROWSERS

South America's biggest mammals, the Amazonian and West Indian manatees (see above), also have poor eyesight but, unlike dolphins, they cannot use sound to find their way. Instead, these barrel-shaped herbivores navigate largely by touch and taste. The Amazonian manatee is a strictly freshwater animal, living in black-water lakes and lagoons, but the West Indian manatee lives in rivers, lagoons, and sheltered areas of the coast. Both have the same unhurried lifestyle, swimming with slow beats of their paddle-shaped tails. Weighing up to 500kg (1,100lb), a manatee can eat up to one-fifth of its weight in vegetation every day. This continuous grazing rapidly wears out its teeth, but manatees have a rare adaptation: their molars are constantly replaced by new ones that erupt at the back of their jaws.

CURIOUS HUMANS

HUMAN IMPACT

Manatees are easily observed from boats, and in some parts of their range – such as Florida – they have become a major tourist attraction. However, their sluggish habits and poor eyesight mean that they are also endangered by shipping. Many bear scars from propellers, and each year, throughout their range in the Americas, several hundred die in collisions with boats.

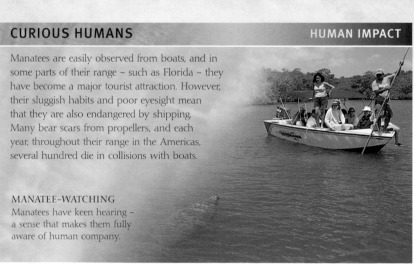

MANATEE-WATCHING
Manatees have keen hearing – a sense that makes them fully aware of human company.

111

TROPICAL RAINFOREST

TEMPERATE WOOD

LOWLAND RAINFOREST

FOREST TYPES
Although South America
has some native conifers, its
forests consist almost entirely
of broadleaved trees. South
America's tropical rainforests
contain a huge variety of
tree species, but its temperate
woodlands are dominated
by southern beeches, which
lose their leaves in winter.
The lowland rainforests are
often flooded during the
wettest part of the year.

FOREST
SOUTH AMERICA'S FORESTS STRETCH FROM
THE ICY SHORES OF SOUTHERN CHILE TO THE STEAMY HEAT OF THE
AMAZON BASIN. HOME TO SOME OF THE WORLD'S MOST ACROBATIC
PRIMATES, THEY CONTAIN A RICHNESS AND VARIETY OF LIFE THAT
HAS FEW EQUALS ANYWHERE ELSE ON EARTH.

FOREST JOURNEY

When European naturalists first ventured into South
America's forests, they were stunned by the scale and
profusion of the plants and animals. In the perpetually
warm and humid climate near the equator, trees grow
to gigantic sizes, and the evergreen canopy creates a
continuous and complex habitat 30–40m (100–130ft)
above the ground. These forests are home to almost
a third of the world's primates (pp.112–17), as well as
to sloths (pp.118–19), tapirs (pp.120–21), and big cats.
They also contain nearly 70 species of marsupials –
successful animals that are often overshadowed by
their better-known Australasian cousins (pp.130–45).

Lowland rainforest makes up the core of this
habitat, forming a belt up to 1,900km (1,180 miles)
wide. Despite recent settlement, the Amazon rainforest
remains one of the largest expanses of forest on Earth.
Unlike forests at higher latitudes, the lowland forest has
an extremely long history, because it persisted during
the last ice age. Consequently, its plants have had time
to evolve into an unparalleled variety, resulting in an
abundance of animal life.

Although the lowland rainforest dominates South
America, many of the continent's mammals live in
different forest types. Some of its rarest primates, for
example, live in the Atlantic forest – a narrow strip
that once stretched almost unbroken down the hills
of the eastern seaboard. Most of this forest has been
cut down, and its primates are now the focus of
efforts to prevent them disappearing from the wild.

NIGHT VISION
EVOLUTION

The world's earliest primates were almost
certainly nocturnal, but almost all of
today's monkeys are active by day. They
have good colour vision, and highly
developed binocularity – vital for judging
distances when jumping. But in South
America, night monkeys, or douroucoulis,
have reversed this trend. They are
nocturnal animals, and their eyes have lost
the repertoire of visual pigments necessary
for colour vision. The most likely theory
for this evolutionary backtracking is that
nocturnal feeding enables night monkeys
to avoid predators, particularly birds of
prey. It also eliminates competition from
other monkey species, which feed by day.

NIGHT MONKEY
Also called the douroucouli, the night
monkey has a small, stocky body
and huge eyes.

Mammals are also common in seasonal forest, a drier type found mainly in Venezuela and northern Brazil. Here, many trees lose their leaves for several months each year, and droughts are common. Mammals found in seasonal forest include jaguars (pp.122–25), cougars (pp.72–73, 102), peccaries (p.102), and armadillos (pp.101, 103).

Finally, some of South America's mammals live in temperate forests, which stretch down from southern Chile to Tierra del Fuego. They include the world's smallest deer (the southern pudu), and also the monito del monte – the sole survivor of a marsupial order that is otherwise extinct.

FIFTH LIMB
Black spider monkeys have a highly effective prehensile tail, which they use as an extra limb to hang from branches. The flexible tip is strong enough to support their body weight.

ENDANGERED TAMARINS

Many primates have subdued colours, but the reverse is true of tamarins. Some species of these small, squirrel-like monkeys, which are found only in South America, are among the most beautiful and striking primates in the world. The golden lion tamarin (see left) is the most eyecatching of all, with a lustrous coat and flamboyant golden-red mane. Like all tamarins, it has claws rather than nails, and it is diurnal, feeding on small animals, flowers, and fruit, and also on gum exuded by trees.

The classification of tamarins is an area of ongoing research. About 16 species are currently recognized (see box, below), but because tamarin populations are often highly localized, many researchers believe that the actual total is higher still. The most recently discovered species, the black-faced lion tamarin, was found in a small pocket of Brazil's Atlantic forest in early 1990. Unfortunately, this forest lies in one of the most densely populated parts of South America, and only 7 per cent of its original extent now remains. The black-faced lion tamarin's population is estimated at just 300, which makes it one of the rarest primates in the world. The golden lion tamarin also comes from Brazil's Atlantic forest and has been the subject of a major captive breeding programme. As a result, the population numbers are on the increase – from about 200 in the 1970s to about 1,000 today.

TAMARINS | PROFILE

Silky fur and luxurious manes are a characteristic feature of this group of monkeys. Their average length is just 60cm (24in), more than half of which is made up by a long, fluffy tail. Lion tamarins make up a small genus of species, all from Brazil's Atlantic forest.

The cottontop, saddleback, and emperor tamarins belong to a larger genus, containing typical tamarins. There are 12 or more species of these, scattered across South America's rainforest zone. They reach as far north as Colombia, and extend eastwards to Peru.

COTTONTOP TAMARIN   **SADDLEBACK TAMARIN**   **EMPEROR TAMARIN**

GOLDEN LION TAMARIN
Emblematic of South American conservation, this beautiful monkey feeds low in the forest understorey, where vines and lianas provide cover and a supply of food.

FOREST-FLOOR MAMMALS

Unlike temperate forests, tropical forests usually have a layered structure, creating parallel habitats that are used by different mammals. The lowest layer is the ground itself. In South America's forests, this is home to a wide variety of terrestrial species, including agoutis (see right) and other large rodents – a particular feature of this part of the world. Agoutis are active by day, and they have a wide-ranging vegetarian diet. They often follow monkeys as they move about overhead, collecting fruit and seeds that the monkeys drop while they are feeding. Monkey troops are used to this behaviour and pay agoutis little attention. However, they react quite differently to other followers: for example, capuchin monkeys (p.116) sometimes shower sticks or urinate on people or dogs on the ground.

Agoutis are unable to climb, but the forest floor is also a base for terrestrial mammals that can get off the ground. One of these is the greater grison (p.121), which hunts birds and small mammals in the lower branches of trees as well as on and below ground. Some rainforest cats behave in a similar way. Ocelots (p.109) and jaguars (pp.122–25) catch most of their prey on the ground, but the much smaller margay (see right) does most of its hunting above the forest floor. It runs at speed along branches, pouncing on unsuspecting squirrels and birds. Uniquely among cats, the hind paws of the margay can swivel around by 180 degrees. Using them as anchors, it is able to climb down trees head-first.

THE UNDERSTOREY

Between the forest floor and the canopy is the understorey – a layer that may start just 5m (16ft) from the ground. This is the zone used by most tamarins (see opposite) and marmosets (p.116), and also by mouse opossums (see right). Mouse possums are South America's most widespread marsupials. True to their name, they resemble mice, but most have prehensile tails, and an opposable big toe that helps them to grip twigs and branches as they climb. Most species are active at dusk and after dark, finding their way around the forest through a combination of smell, touch, and sight. Despite their small size, they can be fearless predators, attacking large rainforest insects as well as eating plant food.

Curiously, many mouse opossums lack the classic marsupial feature – a pouch. Instead, the young hang from the female's teats, looking like a cluster of pink or brown berries attached to her underside. During their development, which can last up to 10 weeks, the female remains remarkably agile and continues to climb and to hunt for food.

THE CANOPY

For most of South America's tree-dwelling mammals, the canopy is the key part of the forest habitat. Here, a maze of branches forms an almost continuous thoroughfare, full of scent-marked paths between one feeding place and another. For leaf-eaters, such as howler monkeys, the canopy gives permanent access to an almost endless source of food. Howlers are the largest monkeys in the Americas, as well as the noisiest. In still air, a male howler's call can be heard at a distance of over 1.6km (1 mile) away – a most effective way of advertising ownership of a treetop feeding territory. There are six species of howler monkeys, and their strictly vegetarian habits have far-reaching effects on the way they live. Unlike other monkeys, they tend to be slow and sluggish, and they spend at least half their time resting or asleep. This apparent laziness is a direct result of their diet, because leaves are a low-energy food and take a long time to digest. However, this is not a problem as leaves are everywhere in the canopy, so howlers rarely have to venture far to eat.

Howler monkeys share the canopy with a wide range of more active monkeys that feed on buds, fruits, and seeds. Some monkeys also eat animal food – for example, sakis (see left) sometimes catch mice and birds, tearing them apart with their hands.

TREE PORCUPINE
South America has eight species of tree porcupine. These spine-covered rodents feed on leaves, flowers, and fruit, and many have prehensile tails.

SAKI
Unlike many American monkeys, the saki does not have a prehensile tail. Instead, it often hangs by its hindlimbs to feed.

TREETOP ISLANDS

The highest forest layer is a discontinuous one, formed by giant trees known as emergents. In the forests of the Amazon Basin, the trees can reach up to 65m (210ft) high, and they tower over the forest like islands set in a living sea. While they are flowering and fruiting, these trees attract large flocks of bats after dusk, and spider monkeys (p.113) visit them during the day. Spider monkeys are the most agile of all South America's primates. They normally live high in the canopy, and they hardly ever come down to the ground. However, with birds of prey flying overhead on the lookout for food, emergent trees are not safe as long-term homes. As soon as monkeys have finished feeding, they make their way back into the safety of the canopy below.

FOREST DWELLERS

PRIMATES are found in all forest layers. The grey woolly monkey walks on the ground on two legs, using its tail for balance.

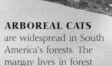

ARBOREAL CATS are widespread in South America's forests. The margay lives in forest habitats from Mexico to as far south as Argentina.

RODENTS live on the forest floor. The agouti relies on its keen senses and speed for self-defence, leaping 2m (6½ft) if startled.

MARSUPIALS in South America are small and mostly arboreal. The common mouse opossum, the most widespread of the group, is an agile climber.

115

SIZE AND FEEDING HABITS

South America does not have any lemurs or bush-babies, and neither does it have any apes. But its monkeys show a great variety of shape and size, with numerous lifestyles and feeding habits to match.

The largest species is the red howler monkey, at about 1.3m (4½ft) long from nose to tail and weighing up to 9kg (20lb). At the other extreme, the pygmy marmoset (see box, right) weighs just 120g (4oz), and is small enough to curl up in a human hand.

Howler monkeys are bulky because leaf eating works most efficiently on a large scale. Their size restricts them to the canopy, because saplings and small climbers cannot bear their weight. The same is true of woolly monkeys (p.115) and spider monkeys (p.113), although these are smaller than howler monkeys and range further to collect their food. Woolly monkeys eat large amounts of leaves, while spider monkeys feed mainly on fruit.

For medium-weight monkeys, which weigh under 5kg (11lb), animal food often makes up an important part of the diet. Compared to howlers, these monkeys are more agile, so better at pursuing prey. For example, capuchins (see below and right) can climb onto thin branches, collecting food, such as vertebrates, insects, and even small bats, that heavier species cannot reach.

Spider monkeys, which weigh just 1kg (about 2lb), are completely omnivorous, picking through foliage to

The smallest of all South American primates, marmosets also have the most specialized lifestyle. Many species feed largely on plant gum, which they collect by gouging holes into the bark of trees and lianas. Unlike tamarins (p.114), marmosets have long incisor teeth in their lower jaw, and they use these to cut out tiny bark chips. Geoffrey's marmoset and the silvery marmoset both live in eastern Brazil, but the pygmy marmoset is from the western Amazon.

GEOFFREY'S MARMOSET

SILVERY MARMOSET

PYGMY MARMOSET

find insects, fruits, and seeds. At the bottom of the weight scale, tamarins (p.114) and marmosets (see box, above) often have a small range, but they eat a variety of specialist foods, such as flowers, nectar, and tree gum. They can cling to vertical tree trunks as they feed, while most larger monkeys need branches or twigs for support.

BROWN CAPUCHIN
This widespread South American monkey uses stone tools to break open nuts.

WHITE-FACED CAPUCHIN
Found from Belize southwards to Colombia, this medium-sized monkey lives in seasonal forest, rainforest, and coastal mangrove swamps.

RED UAKARI
With their red, almost hairless faces, uakaris are instantly recognizable. There are several subspecies with differing fur colours, including white, buff, chestnut brown, and reddish orange.

117

THREE-TOED SLOTH
Clinging to a sapling, a three-toed sloth peers at an unfamiliar environment – open ground. Sloths have two or three claws on each foot. The claws close against a fleshy pad to grip slender branches or trunks.

COPING WITH CLIMATE

For human visitors, the Amazonian climate can be oppressive. The daytime temperature hovers close to 30°C (86°F), and during the wettest months, the forest gets an almost daily drenching with rain. Even when it is not raining, high humidity makes moving uncomfortable. But rainforest mammals appear to take this climate in their stride. Smaller mammals, such as bats, make use of large leaves to keep them cool, while fur, which looks distinctly out of place in a hot and humid climate, is an important asset that works in several ways. It helps the animals to regulate their body temperature and it also keeps them dry. Coloured fur enables them to recognize their kin, but conversely, camouflaged fur helps them to hide away. Tamarins (p.114) and marmosets (p.116) are prime examples of fur being used for recognition, while sloths (see left and opposite) are the unrivalled experts at using fur for camouflage.

Sloths are remarkably successful animals. There are five species in Central and South America, and in many places they consume more food than all other mammalian leaf-eaters combined. Their legendary slowness makes them hard to spot, and so does the strange greenish

> "Even for leaf-eaters, their **metabolic rate** is remarkably low. It can take them a **month** to digest a batch of leaves."

SLOW BEGINNINGS   EVOLUTION

Today's sloths are direct descendants of animals that lived on the ground. One of the earliest species, called *Hapalops*, was just 1m (3ft) long, but later kinds grew to immense sizes. For example, *Megatherium* (see left) was about 6m (20ft) long, and probably weighed up to 3 tonnes (6,600lb). Ground sloths were vegetarians, and they reached their food by climbing, or by rearing up on their hindlimbs and using their hooked claws to pull leafy branches within reach. They walked on their knuckles, with their toes turned upwards, a gait that kept their claws off the ground. Today's sloths have the same kind of claws, but use them to hang upside down.

GROUND SLOTH
Megatherium (left) was large and fed on the ground, but *Hapalops* was small enough to climb up into trees.

tinge of their fur. This green colour is produced by microscopic algae that grow in grooves in the sloth's hairs. The algae use the sloth as a mobile habitat; in return, the sloth gets a highly effective disguise.

Sloth fur has another unusual feature that helps them adapt to the rainforest: the hairs on its trunk run in the "wrong" direction, from its front to its back. As sloths spend most of their lives upside down, this arrangement allows them to shrug off rain.

SLOTH BEHAVIOUR

Because sloths are such unusual mammals, their behaviour has attracted a lot of research. Their diet is almost entirely made up of leaves, but they are highly selective about which ones they eat, feeding on less than one per cent of the trees within their reach, and ignoring all the rest. Even for leaf–eaters, their metabolic rate is remarkably low. It can take them a month to digest a batch of leaves, and food in transit through a sloth's digestive system can make up a third of the animal's weight. A sloth's body temperature rises and falls more than that of most other mammals, dropping sharply when it is inactive – for example, during periods of persistent rain. And while some sloths are nocturnal, most feed at any time, although even the busiest sloths are on the move for just a few hours each day.

One of the strangest aspects of sloth behaviour concerns their visits to the ground. Leaves provide sloths with all the food and water that they need, so in theory they can stay aloft indefinitely, provided the trees are close enough together. However, all sloths climb down to the ground once or twice a week, where they visit habitual "toilet sites" near the trunks of trees. Why sloths should do this is unknown, particularly as they move with difficulty on the ground. This action makes them easy targets for predators, even though they defend themselves by lashing out with their front claws. One possibility is that it is a relic form of territorial behaviour, dating back to an era when ancestral sloths lived partly on the forest floor (see box, opposite).

A QUIET LIFE
Female sloths give birth to a single young. The young sloth clings to its mother's chest for up to six months, long after its diet has changed from milk to leaves.

TENT BATS
Tent bats roost under large leaves, sometimes biting through the supporting struts so that the leaf folds around them as protection from rain.

PRIMATE LOOKALIKES

Sloths are found only in the Americas, and their unique anatomy and behaviour makes them impossible to confuse with any other type of animal. However, the same is not true of the kinkajou – an agile climber with a prehensile tail and grasping feet. Zoologists initially classified it as a primate, and the same mistake is made by local people who confuse the kinkajou, or martilla, with night monkeys (*monos de la noche*). Although the mistake is an understandable one, the kinkajou is actually an arboreal carnivore that belongs to the same family as the raccoons. Close inspection reveals that the kinkajou has clawed toes, rather than ones with nails like a primate, and it also has a long, narrow tongue that it uses for feeding on fruit and flowers.

Raccoons and their relatives – known to zoologists as procyonids – are adept at life in trees. There are 18 species altogether, and South America's contingent includes the crab-eating raccoon, kinkajou, and three kinds of coati, together with much more secretive animals called olingos. These look like kinkajous, but they are more cat-like, and have exceptionally long and bushy tails. Until recently, the red panda was also classified as a member of the raccoon family, but its status has long been a source of debate. Many zoologists now group it with the giant panda, or in a family of its own.

Olingos feed mainly on fruit, but they catch small animals as well. As climbers, these appealing animals are extremely acrobatic, with a squirrel-like ability to run at speed through trees. Despite their shyness, they are also inquisitive and intelligent – in Costa Rica, they have been known to climb down string to reach sugary food put out for hummingbirds.

Another good climber is the greater grison (see opposite), a member of the weasel family that hunts small mammals and birds. Like most of its relatives, it has short legs, and a slender body that allows it to pursue prey in their burrows. But the same low-slung build also enables it to scamper along branches, using its sharp claws to keep a good grip.

TAPIRS

With their thickset bodies, tapering heads, and trunk-like snouts, tapirs make up a small and very distinct family of forest mammals.

KINKAJOU
The kinkajou uses its prehensile tail as an anchor while it feeds, like a monkey. However, it is actually in the raccoon family. Plant food, particularly fruit, forms most of its diet.

"Tapirs have been called 'living fossils' because they have changed hardly at all over the last 35 million years."

A tapir's snout works like a cross between a nose and a hand – with it, a tapir can distinguish edible plants from poisonous ones and grasp leafy stems or fruit. Like other hoofed mammals, tapirs are vegetarians. Their substantial size means that they need at least 30kg (66lb) of food a day. They are important dispersers of seeds because they digest the fleshy parts of fruit but often scatter intact seeds in their dung. Travelling along well-worn paths, often by night, tapirs feed alone, a solitary lifestyle that is interrupted only when females are accompanied by their young.

There are four species of tapir – the Malayan tapir (p.163) occurs in the forests of Southeast Asia, while the remainder are all found in the American tropics. All four species are good runners and climbers, despite weighing up to 300kg (660lb). The mountain tapir, which comes from the Andes, lives at an altitude of up to 4,700m (15,400ft) and can trot up steep slopes with ease. In Central and northern South America, Baird's tapir (see below) also lives in mountainous terrain.

Wherever they live, tapirs are strongly attached to water. The Brazilian tapir (see far right), which is found throughout the Amazon rainforest, often feeds in rivers, using its snout to collect water hyacinths and other floating plants. Bathing and wallowing are often part of the daily routine, but water also acts as a refuge in case of attack. On land, a smooth, streamlined shape helps when making an escape, as does an extra-tough skin. Tapirs are only sparsely haired, but the tough skin on their hindquarters feels like a leathery shell. Protected by this, they can force their way through dense vegetation almost as quickly as a human can run. However, if they are cornered, they fight back, using their teeth. The Malayan tapir is strikingly marked in black and white, while its American relatives are less distinguished, in shades of brown. All tapir calves are dappled with creamy yellow spots and horizontal stripes at first. This notable feature makes the calves look extremely conspicuous in the open, but it is effective camouflage in the broken light of the forest floor. The same adaptation is shown by many wild pigs, animals that share a similar habitat and way of life. Tapirs have just one calf at a time. For its first week of life, it hides in vegetation, but once on its feet, stays close to its mother for a year or so.

BRAZILIAN TAPIR
Like all tapirs, the Brazilian tapir is a good swimmer. It can submerge itself to throw off a determined attacker, or – more often – to keep biting insects away.

GREATER GRISON
This long-bodied predator is expert at catching roosting birds in the lower branches of trees.

BAIRD'S TAPIR
This tapir is the largest terrestrial mammal in Central America. It lives in a range of forested habitats, including mangrove swamps, and has been affected by hunting and deforestation.

JAG
Sou
of s
hu
doc

a n
hu
Am
ma
pa
the
era
the
in
Too
Am

ma
fur
(pp
Ho
sor
by
go
has
for

FO
Wh
mo
the
rel
Re
ma
wi
for
co
are
loo
co

a l
oft
Ag
pa
bu
he
car
hu
ha

AUSTRALASIA

HABITATS OF
AUSTRALASIA

AUSTRALIA IS AN EXTREMELY HARSH, DRY CONTINENT WITH VAST
AREAS OF DESERT, AND SOILS THAT ARE THE MOST IMPOVERISHED
IN THE WORLD. BUT OTHER AREAS OF AUSTRALASIA – INCLUDING
SMALL PARTS OF AUSTRALIA – HAVE VERY DIFFERENT
HABITATS, INCLUDING VARIOUS TYPES OF FOREST
AND GRASSLAND, AND SOME WETLANDS.

PREVIOUS PAGE:
KOALA
Australia's favourite mammal, the
koala, looks cuddly but it can be
short-tempered and aggressive.

ANCIENT ORIGINS

The lands that make up Australasia – Australia, New
Zealand, New Guinea, and various other islands in the
Pacific Ocean – share a common geological history that
goes back to the Cretaceous period (over 70 million
years ago), when they were joined to South America
and Antarctica, forming a supercontinent known as
Gondwana. By about 60 million years ago, the forces
of plate tectonics had begun the process of breaking
up Gondwana into the land masses that we see today.

The first to split off was New Zealand. It probably
took with it its own fauna of mammals, but a series
of environmental crises, including the
submergence of most of the islands
about 35 million years ago, led to their
extinction. New Zealand has only two
indigenous species of mammal, both
of which are bats; all other mammals
have been introduced by humans
over the last 1,500 years.

Australia, Antarctica, and South
America remained closely linked
until 45–35 million years ago. When
they separated, the mammals of these
continents were probably fairly similar,
but as Australia drifted northwards
over the next 23 million years, the
changing climate caused the evolution
of distinctive fauna.

CHANGING
CLIMATE

The whole Gondwanan
supercontinent was
covered in cool temperate
rainforests, similar to
those found today in
parts of Tasmania. After
the breakup of Gondwana,
the northward drift of
Australasia into much
warmer regions, meant
that the continent's climate
became much more arid.

DESERT (PP.132–33)
The arid heart of Australia is
made up of great deserts. Here,
the scarcity of food and water
creates a particularly demanding
environment for mammals.

INDIAN OCEAN

Ti
S

Kim
Pla

Great Sand
Desert

Hamersley Range

Gibson
Deser

Gascoyne River

Murchison River

Nul

Darling Range

S O U

ESSENTIAL INFORMATION

■ **SIZE** The total size of Australasia
covers 9.477 million square km (3.658
million square miles). Mainland Australia
is the largest landmass, at 8.923 million
square km (3.444 million square miles).

■ **NATIONS** Countries with territories
wholly or partly in Australasia include:
Australia, Papua New Guinea, New
Zealand, Indonesia, Solomon Islands,
Vanuatu, and New Caledonia.

■ **CLIMATE** Very variable across region.
Australia is mainly arid, with local areas
of high rainfall, and a range of climate
zones including tropical, subtropical,
warm temperate, and cool temperate.

■ **HIGHEST MOUNTAIN** Mount Carstenz,
New Guinea, at 5,030m (16,500ft); Mount
Kosciuszko, Australia, at 2,229m (7,310ft).

■ **LONGEST RIVER** Murray–Darling,
Australia (3,750km/2,330 miles long).

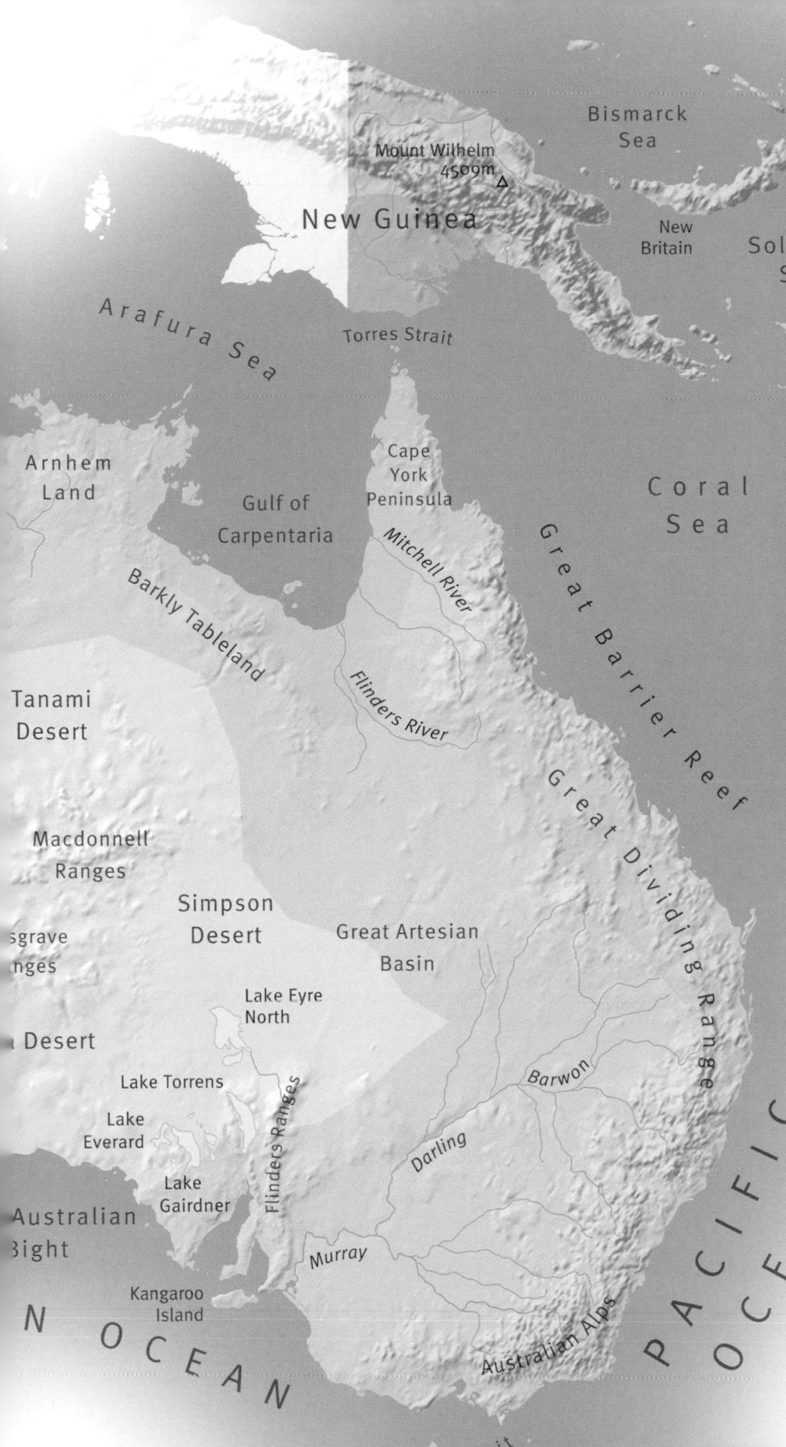

KEY TO HABITATS

- GRASSLAND
- DESERT
- WETLANDS
- FOREST

New Guinea

Mount Wilhelm
4509m △

Bismarck
Sea

Solomon Islands

New Britain

Solomon Sea

New Georgia Islands

Guadalcanal

Santa Cruz Islands

Arafura Sea

Torres Strait

Coral Sea

Vanuatu

Arnhem Land

Gulf of Carpentaria

Cape York Peninsula

Mitchell River

Barkly Tableland

Flinders River

Great Barrier Reef

Great Dividing Range

Tanami Desert

Macdonnell Ranges

...grave ...nges

Simpson Desert

Great Artesian Basin

Lake Eyre North

Barwon

PACIFIC OCEAN

New Caledonia (to France)

... Desert

Lake Torrens

Flinders Ranges

Darling

Lake Everard

Lake Gairdner

Murray

...N OCEAN

Australian Bight

Kangaroo Island

Australian Alps

Bass Strait

Tasmania

Tasman Sea

North Island

NEW ZEALAND

South Island

Southern Alps

Stewart Island

FOREST (PP.136–43)
Australasia was once blanketed in forests. Today, the patchy remains of ancient rainforests are surrounded by newer types of woodland that have evolved to exploit a drier, more seasonal environment.

GRASSLAND (PP.130–31)
Grasses are well suited to a dry environment. For those mammals that can manage to feed on their tough, silica-rich leaves there is an almost limitless food resource.

WETLANDS (PP.134–35)
Even in the driest continent there is water. Heavy monsoon rains generate seasonal swamplands in the northern areas of Australasia, while moisture-rich trade winds supply a network of rivers and streams in the southeast.

ISLAND OUTPOSTS (PP.144–45)
Australia is bracketed to the north and south by the large islands of New Guinea and Tasmania. The ancient forests of these islands provide sanctuary to mammals that are extinct or critically endangered on the mainland.

GRASSLAND IN THE PARTS OF AUSTRALASIA
WHERE THE LAND IS ARID AND THE SOILS ARE NUTRIENT POOR,
MANY PLANTS CANNOT SURVIVE. BUT GRASSES – ONE OF THE MOST
SUCCESSFUL GROUPS OF PLANTS ON EARTH – THRIVE. AS A RESULT,
MANY AUSTRALASIAN MAMMALS HAVE ADAPTED TO GRASSLAND LIFE.

GRASSLAND TYPES
There are many different
grassland habitats in Australia.
Spinifex grassland grows on
rocky ground in arid areas.
Shrubland is a mixture of
grass and bushes, without
a tree layer. The term grassy
or savanna woodland is used
when there are scattered trees,
and the understorey consists
of grass rather than shrubs.

A DIET OF GRASS
Because grasses grow low to
the ground, they are easily
within reach of most animals.
And because they grow from the
base and spread outwards, rather
than upwards from the tip of their
shoots, they can survive cropping;
in fact, the more they are cropped,
the more densely and healthily they
grow. Animals that can eat grass are
guaranteed a self-replacing supply
that can stretch unbroken for hundreds
of square kilometres. However, this
abundant food source carries an
evolutionary cost for grazing mammals.
Grass leaves take up silica from the soil,
an abrasive mineral that wears down
teeth much faster than normal foliage
and makes grass comparatively hard
to digest. Mammals that feed on grass
have had to develop "hypsodont"
teeth, with extra-high crowns, complex
folding patterns of enamel, and
reinforcements of dental cement.

FAT-TAILED DUNNART
This marsupial survives periods
of drought by drawing energy
from the fat deposits that are
stored in its tail.

RED-EARED ANTECHINUS
The male antechinus becomes
extremely aggressive during the
breeding season and invariably
dies after mating.

MARSUPIAL GRAZERS
The main group of native grazing
mammals are the kangaroos. Their
ancestors were small, browsing
mammals that fed on the forest
floor. Close relatives of these, the rat
kangaroos (p.136), still live in Australia's
forests today. The ancestral kangaroos
were small enough to hide in the
undergrowth when danger threatened. As
they shifted to grazing in open grassland,
they began to live in groups for greater
safety. They also increased in size, their
hindfeet became longer and narrower,
their hindlimbs longer and more powerful,
and their tail evolved into a long, heavy
counterbalance, to enable them to cover
long distances on the plains at high speed.
    The other group of marsupial grazers
are the three species of wombat. Like
kangaroos, these bulky, burrowing animals
have high-crowned teeth for eating grass.
However, the wombats have gone one stage
further in coping with the wear caused by
chewing on abrasive grass: their teeth are

PLACENTAL AND MARSUPIAL MAMMALS

EVOLUTION

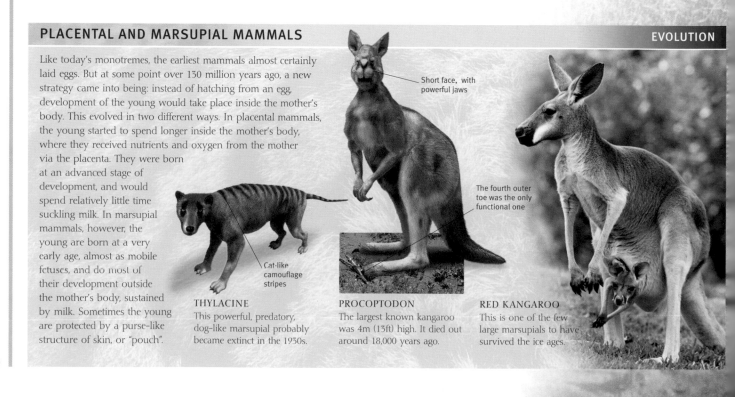

Like today's monotremes, the earliest mammals almost certainly
laid eggs. But at some point over 130 million years ago, a new
strategy came into being: instead of hatching from an egg,
development of the young would take place inside the mother's
body. This evolved in two different ways. In placental mammals,
the young started to spend longer inside the mother's body,
where they received nutrients and oxygen from the mother
via the placenta. They were born
at an advanced stage of
development, and would
spend relatively little time
suckling milk. In marsupial
mammals, however, the
young are born at a very
early age, almost as mobile
fetuses, and do most of
their development outside
the mother's body, sustained
by milk. Sometimes the young
are protected by a purse-like
structure of skin, or "pouch".

Short face, with
powerful jaws

The fourth outer
toe was the only
functional one

Cat-like
camouflage
stripes

THYLACINE
This powerful, predatory,
dog-like marsupial probably
became extinct in the 1930s.

PROCOPTODON
The largest known kangaroo
was 4m (13ft) high. It died out
around 18,000 years ago.

RED KANGAROO
This is one of the few
large marsupials to have
survived the ice ages.

A combination of hunting and setting fire to bush led to the extinction of larger marsupials around 35,000 years ago. When European settlers arrived, they cleared land, displacing or killing many native species, and they introduced mammals (for food, hunting, or as pets) that preyed upon native animals or competed with them for food.

RED FOX
Introduced by Europeans for hunting, foxes have destroyed many of Australia's mammals.

rootless, meaning that they continue to grow throughout the animal's life. As a result, wombat teeth never wear down.

MARSUPIAL CARNIVORES

Australian grasslands lack the spectacular carnivores seen on other continents – there are no lions or cheetahs. Instead, the marsupial carnivores, known as dasyurids, are all small animals. The most abundant are the various groups of marsupial "mice". Unlike true mice, which are rodents and feed on plant material, marsupial mice are hunters, feeding on insects, lizards, and true mice. Although tiny, many of these, such as the broad-footed marsupial mice, or antechinids (see opposite), can be extremely aggressive. In the most arid grassland, some species of narrow-footed marsupial mice, also known as dunnarts (see opposite), build up reserves of fat in their tails so that they can survive long periods of drought.

ANTILOPINE WALLAROO
An inhabitant of Australia's northern grasslands, the antilopine wallaroo is an important game animal for aboriginal people.

"About 20,000 collisions between motor vehicles and kangaroos take place every year."
HOLDEN (AUSTRALIAN SUBSIDIARY OF GENERAL MOTORS)

LIVING IN GROUPS
With their high-crowned teeth, the red kangaroo is able to process even the coarsest grasses. Grouping together is a good way to keep watch for predators in open country.

DESERT TYPES
In Australia, "desert" describes
a range of environments. The
Red Centre is typified by red
sandstones, which can form
spectacular outcrops such as
Ayers Rock and the Olgas.
There are also classical sandy
deserts with dunes. Much of
central Australia consists of
rocky deserts covered in
spinifex scrub.

DESERT

LITTLE RAIN FALLS IN THE CENTRE OF
AUSTRALIA. AS A RESULT, THREE-QUARTERS OF THE LAND IS NOW
DESERT OR SEMI-DESERT. DESPITE THE CONTINENT'S ARIDITY, THERE
IS MORE VEGETATION FOUND IN THESE AREAS THAN IN DESERT
REGIONS ANYWHERE ELSE IN THE WORLD.

IN THE RED CENTRE

Desert presents particular
challenges for mammals,
the greatest of which is the
absence of water. Mammalian
metabolisms require a great deal
of water, far more so than in the case
of cold-blooded animals, such as reptiles.
With little liquid water to drink, desert
mammals must get water from their food
and minimize the moisture that they lose.
In the Australian desert, most mammals
are small and nocturnal. During the day,
they shelter from the heat in burrows that

RED KANGAROO
Symbol of the Australian
bush, the red kangaroo's
hopping gait enables it to
cover vast tracts of open land
both rapidly and efficiently.

maintain a humid atmosphere,
reducing moisture loss from the
body. After dark, when the air is
cool, they come out to forage. Hopping
mice, one of Australia's native rodents,
spend the day in burrows that can run
as deep as 1m (3ft) below the surface. At
night, they emerge to search for seeds and
green plants, which provide them with
both food and water.

The main carnivores of the Australian
desert are small marsupials, such as the
kowari (see box, opposite). Perhaps the
most bizarre of the desert marsupials is

A ROCKY LOOKOUT
Perched high on a rocky outcrop, a pair of rock
wallabies illustrate two advantages of living where
they do: it allows them to see predators coming, and
it provides ample shelter in which to avoid them.

the marsupial mole. Completely blind, this species "swims" through the sand in search of subterranean insects and their larvae. A horny shield protects its snout and, helped by powerful forelimbs with large claws, enables the mole to plough along like a miniature bulldozer. Like many burrowing marsupials, the female's pouch opens to the rear to stop the young from being smothered in sand.

GREATER BILBY
The largest of the desert's marsupial predators, the greater bilby feeds on insects, lizards, and small mammals.

ON THE ROCKS

Rock wallabies (see below) inhabit rocky outcrops and heights in the Australian desert. They are among the most attractive of the kangaroos, with striking fur colours and patterns. These wallabies are ideally adapted for a life leaping around on steep slopes: the surfaces of their feet are extremely rough,

to provide extra grip, and their long, rigid tails provide balance as they jump, sometimes as much as 4m (13ft) horizontally in one bound. In especially dry conditions, they can exist for long periods without water by eating the juicy bark and roots of various plants.

Rocks also provide shelter for the dingo (see right). Dingoes are adaptable animals, hunting alone to catch smaller prey, such as insects and lizards, but banding together in small groups to run down kangaroos. On occasion, they will also attack livestock, which led to their persecution by European settlers. Once found across Australia, dingoes now tend to be confined to arid, less agricultural areas.

TRULY WILD?
Australia's wild dog, the dingo, was actually brought to Australia from Southeast Asia some 5,000 years ago by humans. Dingoes still enjoy a loose association with aborigines, skulking around campsites in search of scraps.

"The animal [rock wallaby] is excessively wild and shy in its habits, frequenting in the daytime the highest and most inaccessible rocks."

SIR GEORGE GREY, SURVEYOR-GENERAL, NEW SOUTH WALES, 1859

SMALL CARNIVOROUS MARSUPIALS   PROFILE

A variety of small but ferocious marsupial carnivores live in Australian deserts. They are all superficially similar in that they have pointed snouts, large eyes and ears, and either lack a pouch or have a pouch that is present only during the breeding season.

The kultar moves like a rabbit, bounding from the long hindlegs on to the fore. The ningaui has a huge appetite, eating almost two-thirds of its bodyweight per night. The kowari stalks its prey like a cat, then pounces and kills with a bite to the neck.

KULTAR   **NINGAUI**   **KOWARI**

WETLAND TYPES
Australian wetlands are permanent in some places, for example the creeks found in rainforests along the eastern coast, or seasonal, such as the flooded swamplands of the Northern Territories.

WETLANDS
LACK OF WATER IS A DEFINING FEATURE OF MUCH OF THE AUSTRALIAN ENVIRONMENT, BUT THIS DOES NOT MEAN THAT THE WHOLE CONTINENT IS DRY YEAR-ROUND. PART OF NORTHERN AUSTRALIA, FOR EXAMPLE, LIES AT THE EDGE OF THE MONSOONAL TROPICS AND ARIDITY AND RAIN ARE SEASONAL.

WATER IN A DRY CONTINENT
During the monsoon season, which lasts from January to March, torrential rains flood the grasslands of Australia's Northern Territories, turning them into vast swamps. Grassland mammals, such as the agile wallaby and the black wallaroo, move to rocky outcrops, feeding on grasses and herbs that grow in the crevices of these temporary islands.

In southeast Australia, the prevailing southeasterly trade winds bring moisture inland from the ocean. When this warm, damp air hits the eastern dividing ranges, it is forced upwards. The air cools, and its capacity to carry water is reduced. The resulting rain falls heavily along the eastern slopes of the mountains, feeding an extensive network of streams, rivers, and creeks. These waterways are the home of Australasia's largest rodent, the beaver rat (see left).

AMPHIBIOUS RAT
The beaver rat, one of Australasia's native rodents, is found only in Australia, Tasmania, and New Guinea. Although most rats can swim, the beaver rat is exceptionally well adapted to life in water. It has a thick, waterproof coat of fur and large, partly webbed hindfeet, which it uses to paddle through the water. Its snout is equipped with sensitive whiskers, which it uses to locate its prey underwater. It feeds on fish, frogs, and crustaceans, as well as large aquatic insects.

The beaver rat does not spend all its time in the water; indeed, in winter it may do most of its hunting on land. It nests in burrows dug into the banks of streams and creeks. These burrows are also used by Australia's other aquatic mammal, the platypus (see box, right, and opposite).

PLATYPUS LIFESTYLE
The duck-billed platypus is a bizarre animal. Its appearance is so odd that the first specimen to reach Europe from Australia, in 1799, was denounced as a fake. Like the echidnas (p.145), the platypus is a monotreme; the female lays two eggs, which she incubates in her burrow by folding them between her belly and her tail. She sits, immobile,

BEAVER RAT
Also known as the Australian water rat, the beaver rat is Australia's heaviest native rodent (it weighs up to 1.25kg/3lb). Its waterproof coat varies in colour from brown to grey, and it has a distinctive, white-tipped tail.

for ten days, exhaling warm, humid air to assist in incubation. Like all mammals, the platypus young suckle milk. But the female platypus has no breasts; instead, milk oozes from the skin of the mother's belly and the young suck it from the belly hairs.

The platypus is the most venomous mammal on Earth. Male platypuses carry spurs on their hindlimbs that are connected to venom glands, and they use these spurs to fight other males during the breeding season. The venom is powerful enough to kill a dog, and causes great pain if injected into a human.

Platypuses are an extremely ancient group. The very earliest mammals probably reproduced in the same way as the platypus, and the oldest known fossil dates back to 110 million years ago – a time when the dominant animals on Earth were the dinosaurs. Like the marsupials, platypuses evolved in Gondwana (p.128) and, except for one fossil found in Argentina, they seem to be confined to Australasia.

UNIQUE ADAPTATIONS
EVOLUTION

The platypus is superbly adapted for life in water (see opposite). To reduce drag, its body is streamlined, flattened top to bottom, and it is covered with dense, short fur. To aid swimming, both the forefeet and hindfeet are webbed. Beneath the skin of its duck-like bill, a special organ is capable of detecting the minute electromagnetic fields that surround the bodies of the worms and small crustaceans on which it feeds. Over the course of its evolution the platypus has lost its teeth; instead, it does its chewing with horny pads on its upper and lower jaws.

PLATYPUS FEEDING
The platypus seeks worms and crustaceans on the stream bed, using its bill to probe among rocks and scoop up sand and gravel.

"A Disbeliever in everything beyond his own **reason**, might **exclaim** [of the platypus], 'Surely two distinct **Creators** must have been at work.'"

CHARLES DARWIN, 1836

PLATYPUS UNDERWATER
Swimming is powered mostly by the paddle–like forefeet, and the hindfeet are used as rudders. The broad, flat tail aids diving and surfacing.

FOREST TYPES
Australia contains tropical and subtropical rainforests, with many species of trees, abundant vines, epiphytic plants, buttressed roots, and a multi-layered canopy. Eucalypt woodland ranges from the towering groves of wet schlerophyll forest to the stunted shrub savanna known as mallee. Temperate rainforests have straight-trunked trees, an even canopy with only two layers, and few, if any, vines or epiphytes.

FOREST
AUSTRALASIA HAS A NUMBER OF DIFFERENT FOREST TYPES, INCLUDING TROPICAL AND SUBTROPICAL RAINFORESTS, WHICH CONFORM TO THE PUBLIC PERCEPTION OF "JUNGLE"; COOL TEMPERATE RAINFOREST; AND EUCALYPT WOODLAND, WHICH IS DRIER AND MORE OPEN THAN RAINFOREST.

AUSTRALIA'S ANCIENT FORESTS

When Australia split away from the supercontinent Gondwana (p.128) some 55–60 million years ago, it was mostly covered in cool temperate rainforests similar to those found in Tasmania (pp.132–33). As Australia moved further north, getting closer to the equator, the forests became more tropical. By 23 million years ago, there were widespread rainforests all over the continent. Today, Australia's rainforests are found only in a thin, broken line down the island's eastern edge. One of the most important areas is along the coast of northeastern Queensland. Here, in the wettest region of Australia (rainfall averages nearly 3m/10ft a year, but sometimes reaches 10m/33ft), grow the last remnants of Australia's tropical rainforests. Although they have been severely depleted by forestry and agriculture in recent years, the australian rainforests retain the highest diversity of animals of any area in the country today. Like the rainforests of New Guinea (pp.144–45), they are a last refuge for mammals that 20 million years ago were found across the continent.

Among the most striking of these mammals are the ringtailed possums. There are 13 species of ringtails, 11 of which are confined to New Guinea and the Queensland rainforests. Unlike most other possums, which will occasionally eat insects or other small animals, ringtails restrict their diet to leaves.

NUMBAT
Numbats use their long tongue to pluck termites from tunnels below the forest floor. They can lick up to several hundred termites per second.

They get their name from their prehensile tail, which is held coiled in a ring when it is not being used as a fifth limb to help the possum move around in the trees. The underside of the tail is naked, which improves its grip. This helps to distinguish the ringtails from the other group of large tropical possums, the cuscuses. Cuscuses have a tail that is completely naked for the last two-thirds of its length. They are mostly found in New Guinea and the surrounding islands (pp.138–39), although there are a limited number of spotted cuscuses (p.140) still clinging on in northeastern Queensland. It is a most spectacular animal, with a vast range of fur colours including pure white, grey with white spots, and bright orange marbled with white. Large, forward-facing eyes help the nocturnal cuscus to judge distances, but it rarely jumps; instead, it moves through the forest canopy at a stately pace like that of a sloth (p.119).

The Queensland rainforests are also the only stronghold of the musky rat kangaroo. This "primitive" marsupial provides us with a glimpse of what the ancestors of today's kangaroos may have looked like. As its name suggests, it looks more like a rodent than a kangaroo; it has a naked, scaly tail, pointed snout, and large, hairless ears. It also runs using all four limbs, rather than hopping on its hindlimbs. It feeds on insects on the forest floor.

EUCALYPT WOODLAND

When the rainforests fragmented as a result of the last ice age, 17,000 years ago, plants developed that were ideally adapted to prosper in the increasingly dry Australian climate. They were the hard-leaved plants (casuarinas, acacias, and the eucalypts) and they formed the nucleus of the "new" woodlands that would displace the rainforests. Eucalypts are particularly well adapted to cope with one of the

SHY BY DAY, BOLD BY NIGHT
BEHAVIOUR

Wild animals do not usually seek out human company, but the rufous rat kangaroo is an exception. By day, it is difficult to approach as it dodges away and seeks shelter in hollow logs. By night, however, it is often attracted to campfires, and may even take food from the hand. It is one of the many species of native Australian mammal to have undergone a dramatic range contraction in the last 200 years. This does not appear to be the result of European-style agriculture (the most common cause of such reduction); the most likely culprit is predation by introduced foxes (p.131).

NIGHT-TIME FEEDER
The rufous rat kangaroo is a rare nocturnal species found in northeastern Queensland.

"Australia's **rainforests** are found **only** in a thin, **broken line** down the island's eastern **edge**."

PROFILE

WALLABIES

There is not much biological difference between wallabies and kangaroos. Generally, the smaller species, such as the three shown below, are known as wallabies, while the larger ones are usually called kangaroos.

SWAMP WALLABY

PARMA WALLABY

QUOKKA WALLABY

major challenges of a dry environment – fire. The tree itself is protected by thick bark, which in a fire usually chars but does not burn. Beneath the bark is an emergency reserve of leaf buds; chemical changes in the bark, caused by the fire, enable these buds to shoot quickly. In addition, the leaves are loaded with flammable oils, which burn only briefly. The main mammal residents of eucalypt woodland are koalas and possums (pp.138 and 139).

There are vast tracts of Australia where even eucalypts struggle to survive. In the areas surrounding the arid interior, fire and drought stunt their growth. Unlike the stately gum trees of the eucalypt woodlands, which have single, straight trunks, these eucalypts have multiple, gnarled, twisted stems. This arid form of growth is known as mallee. In spite of its dry nature, the mallee provides a home for many species of Australian mammal, including bilbies (p.133), sticknest rats, hopping mice, wallabies (see panel, right), wombats, and kangaroos (pp.130–31). But the most unusual mammal found in the mallee is the numbat (see above).

The numbat is the only marsupial that is active exclusively during the day. Its ancestors were carnivores related to the dasyurids and, distantly, to the quolls (p.145) and dunnarts (p.130). The numbat, however, has become a specialist in eating termites. It has 52 teeth – more than any other land mammal – but they are all small and delicate, so it catches its prey using its tongue, which can extend for half its body length. The numbat hunts by day because, unlike the placental anteaters of South America (p.101), it lacks the powerful claws and forelimbs that are needed to rip open rock-hard termite mounds, where termites shelter at night. Hunting by day is risky, and exposes the numbat to predation. Of all the marsupials, it seems to be the one that has suffered most from introduced predators, such as foxes and cats.

SLUGGISH LIFE IN THE TREES

Koalas are largely nocturnal, and almost entirely arboreal. They only occasionally come to the ground, either to shuffle to another tree, or to lick up gravel to help their digestion. Their low metabolic rate leads to a daily behaviour pattern that might best be described as lethargic and comatose; they spend 16–18 hours of each day sleeping, wedged into the forks of trees. Outside the mating season, there is little in the way of social behaviour in koalas. They are mainly solitary, although they sometimes live in loose-knit groups. During the breeding season, koala behaviour changes dramatically (see right).

KOALA FEEDING
Koalas are more or less solitary. Two adults never share the same tree when feeding.

ORIGIN OF KOALAS

The fossil record shows that koalas have been around for at least 25 million years. They are distantly related to the wombats, and the earliest koala was probably a similar sort of animal – a generalized plant-eater that sheltered in a burrow by day and fed on grass and herbage at night. This burrowing ancestry may explain why the koala, unlike the other tree-living marsupials, has a pouch that opens backwards, or upside down. Evidence of the existence of koalas is rarely found in the fossil deposits that preserve the mammal fauna of the ancient rainforests, which may suggest that they had already started to specialize in feeding on eucalypts. Plants of this sort were rare in the rainforests, becoming abundant only when Australia began to dry out during the last ice age.

KOALA DIET

There are very few marsupials that can feed on the leaves of eucalypts. These are tough and leathery, to reduce water loss; they contain oils, which make them indigestible; and they are full of toxins, making them poisonous to many mammals. There are some marsupials that include eucalypt leaves as part of a broader diet, but the koala is the only animal that is capable of living exclusively off the foliage.

Koalas have a number of adaptations to make eating and digesting eucalypt leaves easier (see box, right). Even so, they still have a problem. Since the leaves are poor in nutrients, koalas need to eat as many as possible. But if they consume an excess, the toxins may overwhelm the liver and poison the animal. The koala copes with this dilemma by reducing its energy expenditure and leading a generally inactive life. In addition, it has reduced the size of its brain to conserve energy. The koala is the only mammal to possess such a shrunken brain (a koala's brain does not even fill the cranial cavity).

SPOTTED CUSCUS
The slow-moving spotted cuscus lives in a small area of rainforest in northeastern Queensland and New Guinea.

KOALA BEHAVIOUR

When, during the mating season, koalas rouse themselves from their general state of lethargy, they can become very aggressive. Koala society revolves around dominant males, who share their territories with up to three females. If another male approaches one of these females during the breeding season, he will be attacked by the dominant male. Bellowing loudly, the males fight with teeth and claws, even knocking their opponent from the tree. Mating itself is a violent and noisy process, in which the male and female may actually come to blows.

POSSUMS AND CUSCUSES

Other eucalypt woodland residents include the possums. Originally inhabitants of rainforest, possums needed to evolve special adaptations to cope with the drier, tougher eucalypt woodland. One species that has successfully made this transition is the brushtail possum, which is so well adapted to living outside rainforest that it is often sighted in parks and gardens as well as in eucalypt woodland. The brushtail is a close relative of the cuscuses but differs from them in having a furry tail. (The cuscuses, also known as phalangers, have tails that are hairless from about half way down their length to the tip, see left.)

Wet eucalypt forests in the far southeastern corner of Australia are home to one of the rarest of all marsupials, Leadbeater's possum. This elusive animal was believed to be extinct in 1909 but was rediscovered in 1961.

> "During the **mating** season, **koalas** rouse themselves from their general state of **lethargy**, and can become very **aggressive**."

DIGESTING TOXINS

To cope with its challenging diet of eucalypt leaves, which are toxic in large quantities, the koala has evolved a number of adaptations. Its teeth and jaws are powerful grinders, it has a large hindgut, up to 2.5m (8ft) in length, full of bacteria to extract as many nutrients as possible, and its liver is adapted to filter out eucalypt toxins. To limit the amount of leaves consumed, the koala has reduced its energy needs by being inactive a great deal of the time and by reducing the size of its brain, one of the body's greatest users of energy.

CONSERVING ENERGY
Koalas spend much of their time asleep, start breeding comparatively late (between their second and fourth years), and usually produce only one young a year.

KOALA WITH YOUNG
Just before emerging from their mother's pouch, koala young spend time feeding on her faeces. These contain digestive microorganisms, which may prepare the young koala's gut to cope with a eucalypt diet.

HONEY POSSUM
The honey possum is one of the smallest
marsupials, weighing only 10g (⅓oz), and is
one of the few mammals to feed entirely on
nectar and pollen.

"The honey possum uses its long, pointed **snout** to penetrate the **tube-like** flowers of the banksias; its bristly tongue collects the **pollen** and **nectar**, which is scraped off when the tongue is retracted into the mouth."

FLYING MARSUPIAL
At 1.2–1.5kg (2½–3lb), the greater glider is the largest of the gliding marsupials. It lives in eucalypt woodland and is active at night.

SPACE TO GLIDE

Eucalypt woodlands are more open and lighter than rainforests. Because the trees are more widely spaced, the possums that lived in the canopy had to develop new ways of getting around. When jumping from one tree to another, natural selection favoured those animals that could jump that little bit further. In three groups of possums (the feathertail gliders, lesser gliding possums, and greater gliding possums), folds of skin between the forelimbs and hindlimbs, such as occur in the modern lemuroid possum, grew over time to form gliding membranes. Today, marsupials such as the sugar glider (see box, below) can launch themselves from a high point in a tree and glide for up to 100m (330ft). Just before reaching the target tree, the sugar glider "stalls", dropping its hindquarters so that it hits the trunk with all four feet.

The sugar glider's name is derived from its diet. In drier, more open forests, eucalypt fruits are not sufficiently nutritious. However, eucalypt sap is rich in sugar. The possums feed on the sap, using their prominent incisor teeth to gnaw incisions in the bark, where sap gathers. In species such as the yellow-bellied glider, social interactions have developed because the animals congregate at these feeding trees. More indirectly, possums feed on insects that feed on the sweet sap.

The sugar glider can also access one type of food that other gliding possums cannot digest – the thick, inedible-looking gum of the acacia trees that form the understorey in many eucalypt forests.

> "It [the greater glider]
> ## floats through the air
> in easy, elegant sweeps."
> JOHN GOULD, ARTIST AND NATURALIST, 1863

DIBBLER
The carnivorous dibbler, a rat-sized dasyurid, feeds mainly on insects but supplements its diet with the nectar from banksia flowers.

Sugar gliders have an enlarged caecum, a region of the intestines, that enables them to process otherwise indigestible resins.

HEATHLAND

There are some soils, particularly in the southwest of Western Australia, that are too nutrient-poor even for eucalypts. Other plants take their place, typified by low height, closely packed branches, and small, sharp, stiff leaves. This type of habitat is known as heathland, and the main flowering plants of this habitat are banksias. There are a number of marsupials that feed on banksias, such as the dibbler (see left), and the pygmy possum, a tiny animal that feeds mostly on the pollen and nectar. There is one possum that has taken nectivory to its highest degree. The honey possum (see pp.140–41) is one of the few mammals that depends totally on nectar and pollen for food (the others are all bats). Because it feeds entirely on flowers, it does not need teeth; other than its upper canines and lower incisors, its few teeth are simple, peg-like rudiments. Instead, it uses its long, pointed snout to penetrate the tube-shaped flowers of the banksias; a brush like tongue collects the pollen and nectar, which is then scraped off by ridges on the palate when the tongue is retracted into the mouth.

CLOSE-KNIT GROUPS — BEHAVIOUR

The sugar glider nests in groups of up to six, including adult males and females and their young. All group members (which are descended from a single founding pair) exude a distinctive smell that is recognized by other members of the group, and they will not tolerate strangers in their area. The scents are produced by glands on the head, chest, and genitals in males, and the pouch and genitals in females.

FEEDING TIME
The sugar glider uses its snout to probe for pollen and nectar.

SQUIRREL GLIDER
The folds of skin between the fore- and hindlimbs of the squirrel glider can be extended to form a gliding membrane that will carry it as much as 100m (330ft) between trees.

143

ISLAND TYPES

These contrasting islands – temperate Tasmania and tropical New Guinea – both preserve ancient environments, making them refuges for marsupials that no longer occur in mainland Australia. New Guinea has both tropical rainforest, as in Southeast Asia, and cooler forests at higher altitudes.

ISLAND OUTPOSTS ROUGHLY 160KM (100 MILES) OFF THE NORTH COAST OF AUSTRALIA LIES NEW GUINEA, AND OFF THE SOUTH COAST, ABOUT THE SAME DISTANCE FROM THE MAINLAND, IS TASMANIA. BOTH THESE ISLANDS ARE MOUNTAINOUS, WITH A CLIMATE DOMINATED BY HEAVY RAINFALL.

MARSUPIAL HAVENS

In many ways, New Guinea and Tasmania appear to have little in common. New Guinea is situated in the tropics, and is one of the most geologically active regions on Earth, dominated by earthquakes and active volcanoes. By contrast, Tasmania lies in temperate latitudes, and shares with Australia a remarkable absence of geological activity.

However, there are in fact many similarities between the two. Both islands contain cool temperate rainforests (at high altitudes in New Guinea, and in lowlands in Tasmania) that mimic those that once covered the whole of Australia. As a result, both islands have acted as refuges for types of marsupial that once occurred in Australia but are now extinct on the mainland.

TASMANIAN SURVIVORS

Tasmania is a prime refuge for marsupials, mainly because there are no dingoes on the island (p.135). Tasmania was the last stronghold of the two largest surviving marsupial carnivores: the thylacine (also known as the Tasmanian wolf or Tasmanian tiger, p.130), and the Tasmanian devil (see below). Both these animals became extinct on the mainland

THE JAWS OF A DEVIL

The massive head and jaws of the Tasmanian devil are evidence of its abilities as a scavenger. It feeds on carrion and can consume all parts of the carcass, including fur and bones.

around 3,500 years ago, when the first dingoes arrived. There have been no reliable sightings of the thylacine, a wolf-sized animal with a striped back and a rigid tail, since the 1930s, and it is almost certainly extinct. It was considered a risk to the sheep imported by European settlers, and was shot, trapped, and poisoned indiscriminately. However, the Tasmanian devil is still fairly abundant in some areas of Tasmania.

The island is also the last refuge of the eastern quoll (see right). It was last spotted on the mainland in the 1960s, and is now thought to be extinct there as a result of competition with cats introduced by settlers. It is most common in dry grassland and forest. Tasmania is also a stronghold of the eastern barred bandicoot (p.145), a small marsupial with an elongated head, large ears, and striped hindquarters. It hunts at night for soil-dwelling insects and other invertebrates, which it locates by smell and then excavates with its powerful claws and pointed snout. On the mainland, introduced carnivores, such as cats and foxes, have driven it to the edge of extinction, but it is still common in many parts of Tasmania, particularly in mixed areas of pasture and native bush.

NEW GUINEA WILDLIFE

New Guinea is one of the world's largest and most mountainous islands. It is over 2,400km (1,500 miles) long, with a central chain of mountains. At its highest point, the mountains rise to nearly 5,000m (16,400ft), and are topped by glaciers. This rugged topography has generated a great variety of habitats, including wetlands, dry savanna grassland, hot lowland rainforest, cool temperate montane rainforest, and alpine grassland. Like Australia, New Guinea's native land mammal fauna consists entirely of rodents, marsupials, and bats. Other mammal species, such as pigs, dogs, and deer, have been introduced by humans. New Guinea rodents can reach impressive sizes. There are a number of species of giant rat, the largest of which, the subalpine woolly rat, can weigh as much as 2kg (4lb) and may be the world's biggest rat.

Many of New Guinea's mammals have adapted to life in the forest canopy. Possums make up one of the most diverse groups They range from tiny species, such as the feathertail glider and the sugar glider (p.145), to the larger, sloth-like cuscuses (see left), the biggest of which can weigh up to 7kg (15lb). The most distinctive of the canopy dwellers are the tree kangaroos. Their ancestors left the ground for the treetops millions of years ago, and the adaptations needed for climbing have given them a very different appearance from their ground-living cousins. To help them with climbing, the back feet and hindlimbs of tree kangaroos are much shorter and their forelimbs are more powerful. There is less danger from predators in the canopy, so their eyes and ears have grown smaller, and their snouts are shorter. On the rare occasions that they come down from the trees, they tend to move on all fours more frequently than other kangaroos.

Because New Guinea is so rugged and heavily forested, many areas have only recently begun to be explored by scientists. For this reason, it is one of the few areas of the world where new species of mammals are still being discovered. In the early 1990s, for example, researchers discovered the dingiso, a new species of tree kangaroo. This black and white animal is a tree kangaroo that, bizarrely, has become secondarily adapted to life on the ground.

COMMON CUSCUS
This arboreal omnivore, an agile climber, is found along the north coast of New Guinea, as well as on the islands of the Solomon and Bismarck archipelagoes.

EASTERN QUOLL
Also known as the marsupial cat, this spotted marsupial, found only in Tasmania, hunts at night for native rats and mice as well as introduced rabbits.

"In a state of **confinement** they [Tasmanian devils] appear to be **untameably** savage, biting severely, and uttering at the same time a low, **yelling** growl."

G. P. R. HARRIS, NATURALIST, 1808

EGG-LAYING MAMMALS   EVOLUTION

Like their better-known relative, the duck-billed platypus (pp.136–37), echidnas are egg-laying mammals. However, unlike the platypus echidnas do not use a burrow. Instead, the female develops a temporary pouch, in which a single egg is incubated for about 10 days. After hatching, the young echidna stays in the pouch for about 55 days until it is ejected. There are two species. The short-nosed echidna occurs in most of Australia, as well as Tasmania and central and southern New Guinea. The long-nosed echidna is much larger, and is found only in certain parts of New Guinea. When disturbed, the short-nosed echidna burrows into the ground or rolls up into a spiny ball.

HUNTING FOR TERMITES
The short-nosed echidna ploughs up forest litter with its snout and catches termites with its long, sticky tongue.

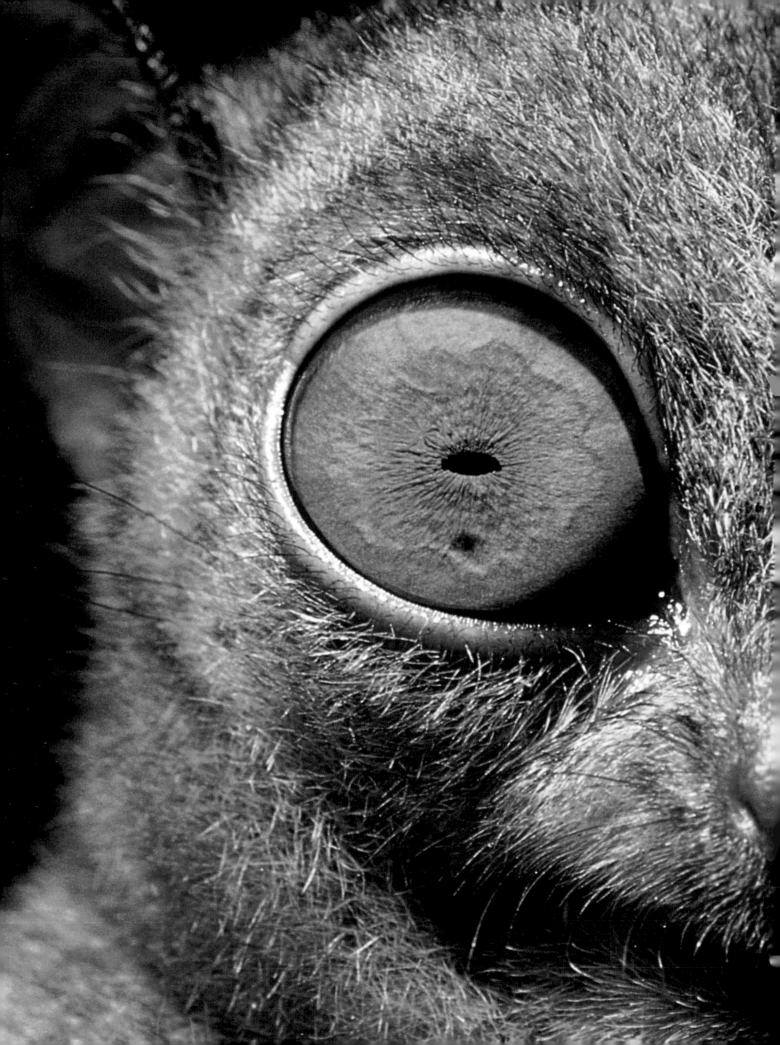

ASIA

50 MILLION YEARS AGO Most regions in what is now mainland Asia, apart from India, have drifted together. However, a seaway separates the western edge of this mainland from Europe. Tropical forest thrives throughout most of the region.

25 MILLION YEARS AGO Having left Antarctica in the far south more than 100 million years earlier, the Indian subcontinent is now colliding with the Asian mainland, and the incredible pressures buckle rocks along the collision zone to create the early Himalayas. Sea levels are falling as the climate cools, exposing land bridges between many islands in the southeast.

10 MILLION YEARS AGO The shape of Asia as we know it today is largely complete (although at this time the Himalayas are less than half of their present height). Ocean-floor trenches between major islands of the east and southeast deepen, reducing the likelihood of a land link when sea levels fall.

1 MILLION YEARS AGO As a result of a particularly severe ice age, the entire northern half of Asia becomes dominated by ice sheets. A new species of mammal is spreading east across the southern continent of Asia – our prehistoric cousin, *Homo erectus*, known informally as Peking Man and Java Man, named after Asian sites where fossilized remains were uncovered.

PREVIOUS PAGE:
TARSIER
These enormous eyes belong to a nocturnal, insect-eating hunter from the forests of Southeast Asia.

HABITATS OF
ASIA

ASIA IS THE LARGEST CONTINENT, FORMING ONE-THIRD OF THE EARTH'S ENTIRE LAND AREA. IT ALSO ENCOMPASSES EVERY TERRESTRIAL AND FRESHWATER HABITAT – AND EXHIBITS EXTREME EXAMPLES OF EACH. IT CONTAINS THE WORLD'S HIGHEST POINT (MOUNT EVEREST) AND ALSO ITS LOWEST (THE DEAD SEA).

ARCTIC TO EQUATOR

Asia spans half the globe in both dimensions, from west to east, and from the Arctic Circle to the equator. Within this enormous area, the climate ranges from freezing cold to searing heat, and the height of the land above sea level varies from low plains and valleys to the highest mountains. As a result, Asia contains almost every known type of terrestrial habitat and almost all major groups of mammals – including many that are rare and some that are unique to the continent.

In the far north lies treeless, frozen tundra, which merges into the world's largest tracts of unbroken forest, consisting mainly of conifer trees, known as boreal forest or taiga. Southwards lie major belts of dry grassland, scrub, and desert, including the enormous Gobi. Towards the southwest, water becomes increasingly scarce until the habitat becomes characterized by the drought-ridden sands of the Arabian Peninsula.

In the southern-central area of the landmass is the world's highest and widest mountain range, the majestic Himalayas. These mountains have acted as a massive barrier to many mammal species: unable to cross this boundary they have evolved in different directions on either side. The Himalayas are also an important habitat in their own right, as home to some of the world's most highly adapted high-altitude species, such as the snow leopard, yak, and various species of wild sheep.

TOWARDS THE TROPICS

South of the Himalayas lies the Indian subcontinent, whose many and varied habitats include the tropical forests that extend southeast into the Indo-China Peninsula and onwards across a patchwork of hundreds of islands, large and small. These islands are "hotbeds" of evolution: their year-round warmth and moisture have encouraged rapid diversity, while their isolation from each other has created some of the highest densities of mammal species in the world.

DESERT AND STEPPE (PP.156–57)
In the centre of Asia is the world's highest desert, the Gobi. Dry steppe grassland stretches from it westwards, while its slightly damper southeast is fringed by parched uplands, where scarce valley lakes provide wildlife with welcome water.

Lak
Lad

Black Sea

Caucasus

Anatolia

Mediterranean
Sea

Euphrates

Tigris

An Nafūd

Zagros

The Gulf

Red Sea

Arabian
Peninsula

Ar Rub' al Kh

Hadhramaut

Gulf of Aden

Socc

GRASSLAND (PP.150–51)
Asian grasslands include subtropical savannas of elephant grass and similar tall species in central India, and smaller grassy patch in scattered clearings throughout the forests of the Indo-China Peninsula. Large herds of hooved, sapling-nibbling mammals he to keep the trees at bay.

ARCTIC OCEAN

Franz Joseph Islands

Novaya Zemlya

Kara Sea

East Siberian Sea

Laptev Sea

Koryak Range

North Siberian Lowland

Ural Mountains

Ob

Yenisey

West Siberian Plain

Ob

Irtysh

Central Siberian Plateau

Lena

Aldan

Lena

Vitim

Kamchatka

Sea of Okhotsk

Sakhalin

Kurile Islands

Steppe Sea

Lake Balkhash

Ozero Zaysan

Angara

Bratsk Reservoir

Altai Mountains

Lake Baikal

Plateau of Mongolia

Great Khingan Range

Amur

Lake Khanka

Sea of Japan

Honshu

Syr Darya

Ili

Tien Shan

Gobi

Yellow River

Yellow Sea

Ryukyu Islands

Karakoram

Tarim He

Takla Makan

Altun Shan

Yellow River

Great Plain of China

East China Sea

Hindu Kush

Plateau of Tibet

Himalayas

Mekong

Sichuan Pendi

Yangtze

Indus

Sutlej

Mount Everest △ 8848m

Brahmaputra

Ganges

Irrawaddy

Salween

Taiwan

Hainan

PACIFIC OCEAN

Thar Desert

Narmada

Godavari

Krishna

Deccan

Bay of Bengal

Korat Plateau

Mekong

South China Sea

Philippines

Andaman Islands

Sri Lanka

Andaman Sea

Gulf of Thailand

Sulu Sea

Nicobar Islands

Celebes Sea

Celebes

New Guinea

INDIAN OCEAN

Sumatra

Borneo

East Indies

Banda Sea

Arafura Sea

Java Sea

Flores Sea

Java

Timor Sea

MOUNTAINS (PP.152–55)

The towering Himalayas descend only partially to the north, where the immense Plateau of Tibet, at 4,600m (15,000ft) high, extends Asia's uplands by more than one million square km (386,000 square miles). The Zagros and Elburz ranges in the southwest rise to well over 5,000m (16,400ft).

FOREST (PP.164–73)

Asia's tropical forests once extended from India to the continent's farthest southeast islands, and rivalled the Amazon Basin for numbers of mammal species. But more than three-quarters of this richest of habitats disappeared in the 20th century alone, and the loss by uncontrolled logging and burning continues at terrifying rates.

KEY TO HABITATS

GRASSLAND

DESERT

MOUNTAINS

WETLANDS

FOREST

WETLANDS (PP.158–63)

Seasonal downpours top up swampy wetlands throughout tropical Asia, especially along low-lying coasts such as the immense Sunderbans at the Ganges-Brahmaputra deltas, in the Bay of Bengal.

ESSENTIAL INFORMATION

■ **SIZE** 44.4 million square km (17.1 million square miles), including the main islands to the east and south, and within the western border of the Ural Mountains to the Black Sea and Red Sea.

■ **NATIONS** Approximately 43 independent states, including the largest by land area (Russia, at 17 million square km/6.59 million square miles), and the two most populated with people (China and India).

■ **CLIMATE** From Siberian Arctic (average annual temperature –15°C/5°F) and Himalayan plateau (5°C/41°F with extensive windchill), to Arabian desert (30°C/86°F).

■ **HIGHEST MOUNTAIN** The world's top 17 peaks are in the Himalayas, the tallest being Everest (8,848m/29,028ft high).

■ **LONGEST RIVER** Chang Jiang (Yangtze), China (6,300km/3,915 miles long).

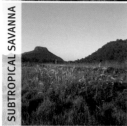

GRASSLAND
ASIA'S IMMENSE DIVERSITY OF GRASSLANDS PEAKS IN CENTRAL-NORTHERN INDIA, WHERE YEAR-ROUND WARMTH AND LOW, PATCHY RAINFALL ENCOURAGE LUSH LOWLAND MEADOWS, DRIER HILLSIDES, AND CLUMPS OF OPEN WOODLAND – IDEAL MIXED HABITATS FOR LARGE HERBIVORES AND THEIR PREDATORS.

GRASSLAND TYPES
Asia's many mountain ranges support varied alpine grasslands, where herbs and small, tussocky shrubs thrive in the short summer, before autumn snowfalls blanket them for the long winter. Temperate and subtropical savanna are found in central and southern Asia respectively. Seasonal monsoons (p.158) flood the low-lying grasslands of the Indo-China peninsula.

SOUTHERN MEADOWS
In Asia, only the vast, dry, thinly grassed plains of the central steppes (pp.156–57) rival the scale of African savannas or American prairies. Smaller mosaics of grassy and wooded habitats dot much of the Indian subcontinent and southeast China, but, as elsewhere in the world, much of this tropical and subtropical savanna has been taken over by "tamed" grass in the form of cereal crops. In the remaining natural areas, the most prominent mammals are the large herbivores, especially the hoofed mammals such as deer, antelopes, and some wild cattle.

HOOFED MAMMALS
Chital, or axis deer (see opposite), eat a variety of grasses and browse in woods, roaming in herds of up to 100. They are fairly small, with a shoulder height of 80–90cm (32–36in), but very swift, dashing for cover at 65km (40 miles) per hour. Their spotted coat and branching, sharp-tipped antlers contrast with the corkscrew-like horns of the blackbuck antelope – smaller, lighter, and having dark brown upper parts and white beneath. In both, the head adornments are multi-purpose, serving to impress rival males at mating time and acting as defence.

Much larger than the blackbuck antelope is the cow-sized nilgai, its cattle-like horns and steely-grey or bluish coat giving rise to its alternative name of blue

SIKA
The sika or Japanese deer is usually rich-red brown with white spots in summer and is almost black in winter.

bull. At breeding time, males compete by lowering their heads or kneeling on their front legs and tossing their horns into each other's faces.

THREATS TO SAFETY
Grassland deer, antelopes, and wild cattle all feed mainly after dawn and before dusk, and rest around midday. They retreat to the cover of tree clumps at night, where they sniff and listen for danger, often in the form of tigers and vengeful farmers.

Another herbivore that is also in danger from humans – even though it is in theory protected by law – is the pygmy hog, which occurs in damper meadows in northeast India. The smallest of all the pigs, at only 60cm (24in) long and 10kg (22lb) in weight, it is poached

HUNTING COMPANION
HUMAN IMPACT

Across Asia (and Africa) impressive predators such as cats have been "tamed", either as showpiece pets or as practical hunting companions. Species include the medium-sized caracal, or southern lynx, and even cheetahs, leopards, and tigers. However the "training" seldom had a lasting effect, and if the cat followed its instincts by trying to escape or defend itself, it was usually killed.

CARACAL
The caracal has narrow, tufted ears and a tawny coat for camouflage in dry grassland. It tackles prey as large as hares or young antelopes.

RAPID REFLEXES   BEHAVIOUR

Battles between mongooses and venomous snakes, such as cobras, are celebrated in many Asian folktales as good (the mongoose) versus evil (the snake). Mongooses rely on their acute, very fast reflexes to avoid a snake's strike.

PARALYSING THE VICTIM
A mongoose bites the top of a snake's neck in order to sever its spinal cord.

for its meat, and in addition has suffered from drainage and the conversion of its grassy riverside habitat into cropland.

FAST DART, SLOW SHUFFLE

Numerous smaller mammals criss-cross the meadows and pastures of southern Asia while foraging. Three species of Indian mongooses (see box, left) – the brown, grey, and small mongooses - are darting, agile hunters of any smaller creatures they find, from worms to snakes. Pangolins (see above) have a very different lifestyle, shuffling along slowly as they lick up ants and termites. They are good climbers, and roll into a ball for all-over protection if in danger.

CHINESE PANGOLIN
The pangolin flicks out its amazingly long tongue up to 40cm (16in) to lick up tiny insect prey. It also has long, curved claws for burrowing.

WATCHFUL AND WARY
Chital study the photographer as they feed, ready to bound off in a split second. Like most large grassland herbivores, they have keen senses and long legs for swift escape.

SPOTS ON THE SLOPES
High up in the Himalayas, wild sheep trek along their autumn trail from summer pasture to lowland shelter. In winter blizzards, these hardy bovids can settle and survive in the open, without food, for several days.

MOUNTAINS
ASIA HAS MORE UPLAND THAN ANYWHERE ELSE IN THE WORLD, DOMINATED BY THE AWESOME HIMALAYAS AND PLATEAU OF TIBET. RELATIVELY FEW MAMMALS SURVIVE HERE, AND THOSE THAT DO ARE HIGHLY ADAPTED TO THE INTENSE COLD AND DRYING WINDS ON THE "ROOF OF THE WORLD".

MONTANE EVERGREEN

ALPINE SCRUB

PEAKS AND PLATEAUX

The Himalayas is by far the world's highest mountain range, with 13 peaks exceeding 8,000m (26,250ft). It is also the widest range, up to 400km (250 miles). Added to this giant habitat of snowy summits, jagged ravines, rocky scrub, and sparsely grassed meadows are the surrounding ranges of the Hindu Kush to the west, Tian Shan in the northwest, and Nan Shan to the northeast. They ring the Plateau of Tibet: more than one million desolate square kilometres (386,000 square miles), average elevation 4,300m (14,100ft), daily temperatures rarely exceeding 20°C (68°F), and only 250mm (10in) of rain each year. In winter, the Himalayan snowline creeps from about 4,000m (13,100ft) down to 1,500m (4,900ft). Even so, tough grasses, mosses, shrubs, and other plants support a variety of mammals, all of which are supremely adapted to the harsh conditions.

HOOFED MAMMALS

Some of the most noticeable mountain residents are the hoofed mammals, especially wild sheep and goats, for example the markhor (see right), the bharal, the Asiatic mouflon, and the tahr (see box, below). They have thick fur coats to protect them against the cold, strong legs for agile leaps among rocks, and wide, splayed hooves with a softer inner pad, to work like "suction cups" on the slippery rocks.

The smallish alpine musk deer shares their habitat. Of its four toes on each foot, the outer ones are unusually strong for a deer and enable better grip, spread the body weight, and minimize sinking in soft snow. Like many mountain dwellers, these herbivores migrate to higher pastures, generally 3,000–4,000m (9,850–13,100ft), for summer growth of grass and herbage, then retreat in autumn to the shelter of more wooded lower slopes.

GORAL
Most goats cope with steep slopes, but gorals of the Himalayas find seemingly invisible footholds on near-vertical cliffs.

They are harried by a variety of large carnivores, from golden jackals (p.156) in the western foothills and Eurasian lynx (p.169) in the north, to dholes (p.165), the ubiquitous grey wolf (p.90), and occasionally the Asiatic black bear (p.166).

PREDATORS AND PREY

The largest purely predatory mammal of the Himalayas is the luxuriously furred snow leopard (p.169). It is slightly smaller than the leopard itself, at 60–75kg (132–165lb), and is a true opportunist with a huge range of prey. It may tackle a yak (p.157), and often overpowers a sheep or goat such as the short-horned goral (see left). In hard times it snatches smaller victims such as hispid or Himalayan hares of the eastern grassy foothills, or a marmot (pp.154–55). The snow leopard may even chance upon an Asian or Himalayan mole, a species similar to the European mole (p.179), on the rare occasion that it emerges from its burrow to locate a mate. Other snow leopard snacks include the smaller-eared cousins of rabbits known as pikas. There are several species, including the Royle's and red pikas. These active, darting, shrill-whistling herbivores prepare for winter by gathering vegetation into a large "haystack" near their den or burrow. On winter excursions from its burrow, the pika consumes its stack, which stays fresh in the dry, cold air.

MOUNTAIN TYPES
In damper climates, such as the southern Himalayas, montane evergreens, such as rhododendrons and tree-sized bamboos, cloak the slopes. To the north there is much less rain, and it runs quickly off the bare slopes and rocky outcrops, so only tough alpine scrub can survive.

MOUNTAIN MONKEY
One of the few Himalayan primates is the furry, snub-nosed monkey of the far northwest.

ASIAN GOATS AND SHEEP
PROFILE

Asian mountain sheep and goats inhabit different and specific regions of the Himalayas. Bharal, or blue sheep, are found in the northwest region; Asiatic mouflon, considered to be ancestors of farm sheep, live in the southwest; while tahr inhabit the highest altitudes, above 5,000m (16,400ft), along the northern fringes of the Himalayas.

BHARAL

ASIATIC MOUFLON

TAHR

DESIRABLE HORNS
Some people regard the markhor's spectacularly spiralling horns as a valuable trophy. This species is also hunted for its meat and hide.

PANORAMIC VIEW
In the Tian Shan Mountains, a long-tailed marmot rises on its haunches to look around. If danger is detected, it dashes to one of its six or more burrows, a complex of tunnels totalling perhaps 100m (330ft).

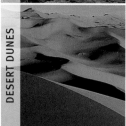

DESERT AND STEPPE IN

CENTRAL AND MIDWEST ASIA, IT IS DIFFICULT TO SEE WHERE STEPPE GRASSES MERGE INTO DESERT SCRUB. BOTH THESE HABITATS ARE SO IMMENSE THEY DEFY THE HORIZON – THE NEAREST TREE COULD BE 200KM (125 MILES) OR MORE AWAY.

DESERT AND STEPPE TYPES

Desert and steppe mammals cope with huge temperature ranges. The Gobi can be 50°C (122°F) at midday in July, and –40°C (–40°F) at midnight in January. Mammals have thick fur to protect against both extremes. Only about one-eighth of the Gobi is sandy with wind-heaped dunes. Much is loose gravel or rocky upland with outcrops of granite.

ARID LAND

At the heart of Asia lies the Gobi, the world's second-largest desert, at 1.3 million square km (500,000 square miles). Its name, meaning "place without water" in Mongolian, could not be more apt. Annual rainfall is a meagre 200mm (8in) in the west, and even less than 80mm (3in) in the east (compare this with New York, which has 1,120mm/44in per annum). The flat, treeless plains are swept by unrelenting dry winds. Little vegetation can survive in the gravelly, often salty soil, exceptions being tough grasses, hardy herbs such as sage, and scattered small nitre, tamarisk, and other bushes. The main mammal groups are large grazing hoofed mammals such as asses, antelopes, gazelles, wild cattle, camels, and small rodents that hide by day in burrows. But these inhabitants are so sparse, larger carnivores must roam dozens of kilometres for the chance of a substantial meal. Many animals have adaptations that help them cope with this region's harsh conditions.

Two perissodactyls eke a living in the Gobi: Przewalski's wild horse (see box, opposite) and the onager, or Asian wild ass. At a glance, these could be confused with each other. However, the horse has a long, profusely hairy tail while the ass has a short-furred tail with a hairy-tufted tip. Both species roam immense distances, in small mare-and-foal bands fringed by stallions, in search of any vegetation they can find. In the breeding season, onager stallions bite and kick each other to establish a temporary territory, which they must possess in order to attract mares for mating.

SAIGA
The saiga's enlarged snout with downward-pointing nostrils may be adapted to regulate body temperature.

HIGHLAND CATTLE

The bull wild yak is one of the biggest of all wild cattle, weighing up to one tonne (2,200lb). But wild yak are extremely rare and keep to the higher slopes, where their skirt-like outer coat, about 60cm (24in) long, plus the soft, close-matted underfur, keep out the bitter wind. Like the horse and

GOLDEN JACKALS
These are found in scrub, semi-desert, and sandy wastes in southwest Asia. Pack adults regurgitate food for cubs.

"prey animals are so sparse that larger carnivores must roam dozens of kilometres for the chance of a substantial meal."

DESERT AND STEPPE DWELLERS

CAMELS, such as the Bactrian camel, have tough lips to cope with thorny desert vegetation.

CATS rarely like to dig, but the sand cat has blunt claws from unearthing small prey such as lizards and rodents.

FOXES, such as Rueppell's fox, hunt mammals and birds, but they can survive on insects, shoots, and grasses.

SMALL SEED-EATERS, like the Mongolian gerbil, do not need to drink, surviving on the moisture in food.

the Bactrian or two-humped camel, the wild yak has been domesticated and is an essential commodity in the lives of Mongolian herdspeople. A smaller bovid of the Kirghiz steppes, in Asia's far west, is the saiga antelope (see opposite). Only male saigas have horns. Small bands gather into larger herds for regular migrations to find fresh grazing. This species is classed as vulnerable, threatened by human hunting for meat and hides, and also for the horns which are ground to powder for oriental medicines.

In addition to these large herbivores are smaller ones, chiefly rodents such as the desert hamster, the Mongolian gerbil, the steppe lemming, and about ten species of jerboas, which hop incredibly quickly across the dusty soil on their massive-footed rear legs. Most plug their burrow entrances by day to keep out predators and extreme heat, and to keep in humidity. At night they emerge to scrabble for plant scraps such as seeds, buds, and leaves.

BACK FROM THE BRINK HUMAN IMPACT

Przewalski's wild horse, known locally as the *takh* or *takhi*, was last truly wild in 1969. However, members from captive-bred herds, in wildlife parks and field stations around the world, are being taken back to Mongolia for reintroduction. The programme began in 1992, with small groups of 15 or so carefully selected horses released every two years, following a period of local captivity to become used to the climate, vegetation, and other herd members. The major release site is Hustain Nuruu, a vast area of mountain steppe. However, interbreeding with free-roaming domestic horses could reduce the genetic purity of the Przewalski species.

STOCKY STEPPE-DWELLER
At about 1m (3ft) tall, this horse is smaller than most domestic breeds. It has a large head, thick neck, no forelock, an upright mane, and shortish legs.

VALUABLE LIVESTOCK
Domestic yak are as little as one-third the size of their wild ancestors. They vary in colour and number more than 12 million. Wild yak are classed as vulnerable by the IUCN, with a total population of perhaps 10,000. They are dark brown to black in colour.

WETLAND TYPES
In the lowest, dampest regions of Southeast Asia, large pools and lakes persist all year, fringed by reeds, rushes, and trees. On higher ground, monsoons bring rushing rivers that most creatures avoid. Man-made drainage channels such as ditches extend wetland habitats, mimicking natural marshes and swamps, with patches of drier ground between.

WETLANDS
NO PLACES ON EARTH ARE WETTER THAN SOUTHERN AND SOUTHEAST ASIA DURING THE MONSOON SEASON. HOWEVER, JUST A FEW MONTHS LATER THE SWAMPS AND MARSHES ARE REDUCED TO REMNANT POOLS SURROUNDED BY DRY, CRACKED MUD.

SEASONAL CHANGES
It is hard to comprehend the intensity of seasonal downpours during tropical Asia's monsoons, which peak from about June to September. The chief monsoon zones are along India's southwest coast, and from the Himalayas' southeastern slopes south to the lowlands around the Bay of Bengal. Annual rainfall of 50m (160ft) is common, and some regions exceed 100m (330ft). Most of this rainfall occurs in the rainy season, when the wetlands flood. Between December and March, temperatures remain high, at 25–30°C (77–86°F), yet there is very little rainfall (only 5 per cent of the annual total), so the wetlands dry out.

Into Southeast Asia, heavy rains are more evenly spread through the months, so lowland swamps and marshes keep their creeks and pools all year. The water teems with fish, frogs, crayfish, and other aquatic life. These are meals for highly adapted mammal predators, such as otters (pp.160–61) and desmans (p.163), and several kinds of smaller cats that (unusually for their group) do not mind getting their feet wet. Enormous herbivores, for example elephants (pp.164–65), rhinoceroses (pp.162–63), and buffaloes (see box, opposite), visit the pools to drink, bathe, and wallow, as do large predators such as tigers (pp.168–69) and leopards (pp.168–69) – two of the big cats (together with South America's jaguar, pp.124–25) most comfortable in water.

Several kinds of deer and antelopes also live in Asia's tropical wetlands. The largest is the sambar (see below) – the stag has enormous antlers weighing

SWAMP OR REED CAT
Also known as the jungle cat, this species swims powerfully after fish and frogs.

approximately one-third of a tonne (730lb), each antler growing up 1.2m (4ft) long. The size and long legs of sambars allow them to wade in water 1m (3ft) deep for soft vegetation, and they can easily swim across fast-flowing rivers. At the other end of the size-spectrum is the Indian spotted chevrotain, or mouse deer, which is hardly larger than a pet cat.

Other hoofed mammals that are attracted by lush water plants include the chousinga, or four-horned antelope, which whistles and barks to stay in contact with the herd as it grazes. The male's two pairs of horns are unique among the 140 bovid species. Another member of this family is the Asian water buffalo (see box, right), found mainly in the northeast of India around the Bay of Bengal. Its massive horn span may exceed 2m (6ft), the largest of any bovid, and a full-grown bull weighs over a tonne (2,200lb). Now very scarce in the wild, this species has been domesticated into many, much smaller breeds.

MIDDAY WALLOWERS

On most days, Asian water buffaloes escape the midday heat by bathing in pools or wallowing in mud-holes, almost submerged with just their faces showing. Many other large mammals, even tigers, follow this habit, and so large animal activity almost comes to a standstill in the stifling warmth in the middle of the day. The water or mud bath is cooling and protects the skin from the sun's fierce burning rays. It also keeps away biting flies and other airborne pests, and washes away fleas, ticks, and other skin parasites.

LISTENING OUT
The water buffalo usually keeps its head above water but submerges fully if it senses danger.

WETLAND DWELLERS

GRAZING DEER, such as the sambar, feed on succulent water plants and keep watch for ripples that could indicate a crocodile.

SEMI-AQUATIC CATS, such as the flat-headed cat, have extra-sharp front teeth to grip struggling fish and crush crayfish.

MOONRATS, found only in Asia., are strong-swimming insectivores related to hedgehogs.

ALONE IN DAYLIGHT
Sambar are unusual among deer in being mainly solitary, apart from a female with her fawn. Their hooves are wide to prevent them from sinking in soft mud.

OTTERS

Asia's tropical swamps are eerily quiet at midday, as the birds rest from calling and monkeys cease their chatter. But a gentle "plop" and a small ripple spreading from the bank across a pool show that some predators are in fact active. Among the mammals most highly adapted to both land and water are the otters, with four of the 13 species found in Asia. The most common otter is also the most widespread species, the Eurasian otter (p.183), found over seven-eighths of the Asian landmass. But the smallest of all otters is found only in Asia – the oriental short-clawed otter (see opposite). It is only half the weight of the Eurasian otter, and is found across Southeast Asia's mainland and on larger islands.

Most otters catch their prey in the water or on the bank. All four feet have webbed toes, for swimming, which are tipped by claws for handling prey, scratching in self-defence, and coat-grooming. When swimming,

the rear feet provide most of the propulsion, the front feet are held up to the chest, and the tail is used as a rudder. Otters keep their eyes wide open when submerged, and the profuse whiskers help to locate prey by feel in muddy water or on a rare nocturnal hunt. On land, otters trot with the undulating, sinuous motion typical of their group – the long-bodied, short-legged carnivores known as mustelids. Other members of this diverse group include stoats, skunks, weasels, polecats, minks, ferrets, martens, and badgers.

The favourite foods of most otter species are fish, supplemented by frogs, water-frequenting snakes and lizards, water birds, their eggs and chicks from surface nests, and the occasional aquatic mammal such as a young water vole. However, the oriental short-clawed otter prefers hard-shelled meals such as freshwater mussels, crayfish, crabs, and water-snails. This species is well named, since its stubby claws do not protrude beyond the fleshy end pads of its partly webbed toes.

FISHING CAT
This species frequently hunts in water and has a fast, powerful, grabbing bite. Many other aquatic hunters have long, slim, pointed teeth – an adaptation for grasping slippery prey such as fish.

UNDERWATER BULLET
The oriental short-clawed otter's tapering snout and long, slim body slip through water with minimum resistance. It spreads oils from its skin glands through its thick coat to make it waterproof.

"When **swimming**, the otter's rear feet provide most of the **propulsion**, the front **feet** are held up to the chest, and the tail is used as a **rudder**."

SHORT-CLAWED OTTERS
These otters form groups of 10-12, containing some strong male-female partnerships. They communicate by yips, barks, and chattering noises.

RIVALS ON THE RIVERBANK

An otter hunting along a marshy bank may encounter other mammals also searching for fish, frogs, and similar prey, including various species of monkeys. Many primates avoid water, but several monkeys specialize in a semi-aquatic lifestyle. In southern Asia they include the very common crab-eating macaque (see box, right), the toque macaque of Sri Lanka, and several species of langurs or leaf monkeys, such as the purple-faced, silvered, and capped langurs. As darkness falls, and the otters and monkeys return to their dens and trees, small cats arrive to prowl along the water's edge. The swamp or reed cat, the flat-headed cat (p.159), and the fishing cat (see opposite) are all proficient swimmers. The fishing cat, in particular, can submerge completely to catch a water bird from below, or drop from a low branch onto a fish just under the surface. This cat also scoops up prey while watching from the bank, and manages to crunch shellfish, water beetles, and other hard-cased prey. Compared to other cats, the fishing cat has a longer body, and shorter legs and tail. These are all adaptations for swimming, as seen in more exaggerated form in the otters.

FEELING FOR FOOD
BEHAVIOUR

In marshes and swamps, monkeys both look and feel in the water, running their hands through the mud for hidden foods such as worms and aquatic insects. Usually they stay on the bank or in the shallows, because dangers such as large snakes and crocodiles lurk in deeper pools. Several monkey species have been seen to smash shellfish, crabs, and similar tough-shelled prey onto rocks or logs - one of many examples of tool-use among primates.

CLEAN MEALS
Monkeys such as the crab-eating macaque often wash their food items (even those found on land) free of mud and silt before consuming them.

161

ON THE BRINK
Rhinoceroses across Asia, such as this Sumatran rhinoceros, are highly threatened by poaching. Wallowing in mud is an important daily activity to keep the rhino's very thick, nearly hairless skin supple and crack-free.

MANGROVE MONKEYS

Many Asian wetlands, especially around the Bay of Bengal and the major islands of Southeast Asia, extend through lowlands to the coast. Here, fresh water becomes salty, and mangroves and other trees grow in thick mud. Mangroves also line the banks of sluggish rivers inland. They present a specialized habitat with relatively few mammal species. One of the most distinctive is the proboscis monkey, named for its pendulous nose. The nose is relatively snubbed in youngsters, grows straighter and more pointed in adult females, and longer and droopier in adult older males, where it may even hang below the chin.

Proboscis monkeys are among the few mammals to survive in mangrove trees, which have leaves and fruits that are relatively low in nutrients and difficult to digest. The monkeys search for trees loaded with seeds, unripe fruits, or young, soft leaves. But such a nutrient-poor diet means many hours of eating, and after a meal more than one-third of the monkey's body weight is the stomach contents, which ferments slowly under the action of microbes in the stomach and gut. Proboscis monkeys soon eat their way through a clump of trees and must then wade across creeks, or even swim rivers by kicking their slightly webbed feet, to find new food sources.

LARGER HERBIVORES

Larger terrestrial herbivores of tropical Asian wetlands include three perissodactyls. One is the Malayan tapir, instantly recognized by its black and white coloration (see box, below). Although it resembles a large wild pig, growing to a weight of half a tonne (1,100lb) it is a closer relation of rhinoceroses and horses. This tapir is seldom far from water. It swims well and frequents muddy wallows and pools, where it rests during midday with just its elongated, trunk-like snout above the surface. At twilight it uses its trunk to probe in soil and mud for fallen fruits, and to grasp young leaves from low vegetation, all the time sniffing so that its extremely keen sense of smell can detect any signs of danger from predators.

Other, much larger wetland perissodactyls are the rhinoceroses. Southern and Southeast Asia are home to three out of the world's five rhinoceros species: two of these (the Sumatran and Javan rhinoceroses) live in wetlands, and there is one grassland inhabitant (the Indian rhinoceros). All five species are threatened and protected. The plights of the Javan and Sumatran rhinoceroses (see opposite), could not be more critical. The Sumatran numbers just a few hundred, and there are probably fewer than one hundred Javan rhinoceroses remaining, in two remnant populations, one in western Java itself and one on the Asian mainland to the north, on the Cambodia-Vietnam border. This species has very broad feet and prefers swampy lowland forests with bamboo, as well as coastal mangroves in Java. Poaching for the horn, which is ground to powder for oriental medicines, is a severe threat to all rhinoceroses. The Javan species has also been decimated as its coastal lowland homes have been logged for timber and then drained for farmland or shoreline development, from industrial ports to tourist centres. Drainage, water diversion, and industrial and agricultural pollution threaten dozens of other Asian wetland mammals, from the great rhinoceroses of Southeast Asia, to small insectivores such as desmans (see above) and water shrews in the northwest.

TROOP LEADER
Proboscis monkeys live in well-ordered troops of up to ten members, consisting of one leading, long-nosed male (above), and several females with their young. At night, several troops may gather in a riverside thicket to sleep, but they rarely trespass into another troop's tree.

RUSSIAN DESMAN
Diversion of rivers to irrigate farmland endangers many Asian wetland species, such as the mole-like desman.

CRYPTIC COLORATION
EVOLUTION

Malayan tapirs have cryptic coloration to break up their bulky body outline. Their coloration alters with age to suit their changing lifestyles. Young tapirs, which mostly lie in dense undergrowth both day and night, waiting to be fed, have brownish coats with pale stripes and spots. The more mobile adults have two-tone camouflage coloration, making them hard to see in moonlight and dark shadows.

ADULT TAPIR
At about 6 months old the tapir's camouflage coloration turns black and white.

YOUNG TAPIR

FOREST ASIA CONTAINS VIRTUALLY ALL THE MAJOR
TYPES OF FOREST, INCLUDING BOREAL OR CONIFEROUS FORESTS
(KNOWN AS *TAIGA*), TEMPERATE AND MONTANE FORESTS, AND SOME
TROPICAL FORESTS, WHICH STILL REMAIN IN SOUTHERN AND
SOUTHEAST ASIA.

FOREST TYPES
Cool, dense, often-dripping
cloud forests of mixed
evergreen broadleaved trees
and conifers cloak valleys and
ravines along the monsoon-
watered eastern Himalayan
foothills. They contrast with
scattered, open groves of
deciduous trees adapted to the
hot, semi-arid, drought-prone
climate of the Maharashtra
Plateau in west-central India.

A DIMINISHING REFUGE
Most of southern and Southeast Asia was once covered
in forest, where heat and moisture provided the ideal
conditions to encourage mammal diversity. But human
activity has had a profound impact on this habitat's
wildlife. Asia's tangled jungles were among the first
tropical forests to suffer from large-scale human
interference, as ancient civilizations cleared them
for homes and farms. In the past two centuries,
disturbance on a far greater scale has taken place –
both directly, by humans hunting and poaching
animals, and indirectly, through habitat loss. Several
of the world's most threatened large mammals, with

populations probably below 1,000, include the Sumatran
and Javan rhinos (see pp.162–63), the anoas of Sulawesi,
and the kouprey (Cambodian forest ox).

Forests throughout Asia have characteristic mammal
inhabitants in each layer, from the canopy down to the
forest floor and beneath, and most typical tree-adapted
mammal groups
are represented,
including
monkeys, such as
the stump-tailed
and bonnet
macaques, and

DAILY BATH
Asian elephants still roam wild in 13
nations, but there may be fewer than
30,000 and their numbers are falling.
The herd's matriarch leads others to
water to drink, bathe, and cool down.

WILD CATTLE
Asian tropical forests shelter
the largest and the smallest wild
cattle, from the gaur or Indian
bison (above), weighing up to a
tonne (2,200lb), to the lowland
anoa (above right), which
can weigh as little as
150kg (330lb).

SOCIAL DHOLES — BEHAVIOUR

Also known as Asian red dogs or wild dogs, dholes are found in Southeast Asia and are a smaller, red-haired version of the grey wolf. They form highly social packs of 20 or more. Dholes are persecuted in some areas for attacking livestock, but are reared elsewhere as semi-tame hunting companions.

HUNTING IN PACKS
Dholes are able to bring down large prey such as deer. All adults within a pack regurgitate food for each other's pups.

numerous outsized climbing rodents, for example smooth-tailed giant tree rats, giant Sunda rats, and Indian giant squirrels, which are pursued through the branches by large cats such as clouded leopards (the smallest and most arboreal of the "big cats"), marbled cats, and other smaller cats.

Gliding is a popular way to escape from predators in Asian forests, and the most expert of all mammals are the colugos, misleadingly called flying lemurs. In other gliding mammals, such as flying squirrels and the marsupials known as gliders, a flap or membrane of skin extends from each side of the body, from front to rear leg. But the colugo's membranes also extend behind the rear limbs to the tail. When the legs are outstretched, the membranes pull taut to form a kite-like structure.

> "In the last few **decades**, at least **50 per cent** of India's forests have been cut down.... The nation's **Biodiversity bill** aims to help maintain **ecological balance**."
>
> INDIAN GOVERNMENT PRESS RELEASE, 2002

FOREST DWELLERS

MANY RODENTS are arboreal. The giant flying squirrel has a gliding membrane to enable it to "fly".

NOCTURNAL CIVETS, such as the oriental linsang, often have large eyes suited to night vision.

TREE SHREWS (here an Indian tree shrew) are not true shrews. They forage by day on branches for insects and grubs.

TREE-DWELLING MUSTELIDS, such as yellow-throated martens, feed on roosting birds.

MOST CIVETS are terrestrial, but the large, shaggy black omnivorous binturong lives in the trees.

BEARS IN THE WOODS

Bears are among the largest, strongest, and most impressive of all forest mammals. Asia has more than its fair share of bears – indeed, six of the eight species are found there. The massive brown bear roams a variety of habitats through the centre of the continent from west to east; the huge, sleek polar bear ranges along the northern coasts; and the remaining four species inhabit the tropical forests of southern and Southeast Asia. These include the sloth bear in the Indian subcontinent, the Asian black bear found through the Himalayas and into the western Indo-China Peninsula, the sun bear, which occurs in the Malay Peninsula and large islands such as Sumatra and Borneo, and the giant panda, which lives exclusively in forests in China.

THE BEAR FAMILY

The bear family (Ursidae) is a very conservative group of mammals, meaning that the various species have diversified relatively little from their common ancestors. Although the different types of bear vary somewhat in size, their overall stockiness and powerful form, their detailed anatomy, and even their general fur colour (except, of course, for that of the polar bear) are remarkably similar to each other. The giant panda's striking black and white coloration, and details of its teeth (it has two additional molars) and other body parts, seem to set it apart from the rest of the bear family. Indeed, for many years it was included in a separate group entirely, the Ailuridae, together with just one other species, the smaller and more fox-like red or lesser panda (see right). But genetic studies comparing DNA from all these mammals showed that the giant panda in fact merits full inclusion in the bear family.

TREE-CLIMBING DOG

Is it a dog, a raccoon, or a bear? The highly unusual raccoon dog is in fact a member of the dog family (Canidae), yet it also displays features of the other groups. It is well suited to life in the forest, since it has highly flexible limb joints, allowing it to splay its wide-set, powerful legs out sideways to "hug" a tree trunk when climbing, like a bear. It also passes the cold season in semi-hibernation. At 7kg (15lb), it rivals the bush dog of Central–South America as the smallest wild dog.

NOTHING REFUSED
Few mammals rival the raccoon dog's omnivorous diet. It eats virtually anything, from birds, mice, fruits, and fish to hard-cased nuts and crabs.

RED OR LESSER PANDA
An able climber with partly retractable claws, this rare and widely omnivorous member of the carnivore order inhabits dense forests at 2,000–4,000m (6,600–13,100ft) altitude, mainly in western China.

FEEDING HABITS OF BEARS

Bears have immensely adaptable diets, and when hungry will not spurn the rotting meat left on an old carcass. But in general, apart from the predominantly carnivorous polar bears, they are less actively predatory, and more herbivorous, than usually imagined. The sloth bear likes fruits and honey, although it also eats "meat" in miniature form, sucking ants and termites noisily through its pursed lips. The sun bear climbs exceptionally well and forages in the trees for fruits, flowers, and varied plants, as well as grubs, honey, and warm snacks such as birds and small mammals. The Asiatic black bear also spends much of its time in trees, searching for soft buds, leaves, nuts, and fruits, plus the occasional grub or bird's egg. These bears' preferences for juicy vegetation have led to conflict across southern Asia as farmers trap, poison, or shoot individuals that leave the forest to raid crops. In southern Asia, the Asiatic black bear is designated as vulnerable and the sloth bear, sun bear, and giant panda as endangered.

THE BAMBOO-FEEDERS

The giant panda is the world's rarest bear, and is the most restricted in range and habitat, being confined to pockets of bamboo-rich, upland forests in central to southwest China. It is also the most extreme herbivore of the family. More than 98 per cent of its diet consists of the soft stems, shoots, leaves, and other parts of bamboo. It manipulates these with what looks like a bendable "sixth finger" on each forepaw. This is really the extension of a bone in the wrist, called the radial sesamoid.

Occasionally, the giant panda diversifies to other shoots and herbs, and also sporadically consumes eggs, grubs, or carrion. These alternatives become more important when local bamboo dies back for a time after flowering and setting seed, every 50 to 100 years. However, expansion of homes and farms have hemmed the pandas ever more tightly into their reserves and sanctuaries, so they cannot forage widely for other foods when the bamboos recede.

BEARS OF SOUTHERN ASIA

The three main bear species of southern Asia share similar physical characteristics in that they all have dark coats and white or pale chest markings. The sloth bear has the shaggiest fur and weighs around 150kg (330lb), the sun bear is the smallest of all bears, weighing about 60kg (132lb), and the Asiatic black bear is the largest, weighing up to 200kg (440lb). The Asiatic black bear's white patch on its chest gives it the alternative name of moon bear.

SLOTH BEAR

SUN BEAR

ASIATIC BLACK BEAR

"Local people call it bei-shung, or white bear....it is a rare beast and shy when approached, preferring to shuffle quietly among tall stems."

PERE DAVID, 1826–1900

RARE BEAR
There are an estimated 1,000 giant pandas in the wild. There are major concerns about inbreeding, since many panda reserves are too small to hold enough individuals for sufficient genetic variety.

TYPES OF TIGERS

Tigers once spanned the whole of southern Asia, from the Caspian Sea to the southeastern island of Bali. They evolved across this vast range into eight subspecies. Three of these are now extinct, including the Javan tiger, which became extinct as recently as the 1980s. The rarest subspecies living today is the Amoy (South China) tiger, with only 30 individuals. Some conservationists feel it is already doomed.

SUMATRAN TIGER
This is smallest surviving subspecies and it also has the darkest coloration. Only 400–500 remain.

ON THE PROWL

Tropical Asia's greatest land predator is also the largest big cat – the tiger. The biggest subspecies is the Siberian tiger, which lives not in Siberia proper, but in the snowy Manchurian Plain of northeast China and southeast Russia. This magnificent hunter may achieve a head–body length of 3m (10ft) and weigh 300kg (660lb). Its estimated numbers are approximately 300, of a total world tiger population of 5,000–7,000. Two-thirds of these are Bengal tigers from the Indian subcontinent, where they keep mainly to thick cover in forests and tangled scrub covering ravine-scarred country. The instantly recognizable black stripes on

a yellowish coat provide excellent camouflage at dusk, as the tiger sets off on a prowl that can extend to 20km (over 12 miles). The standard hunting technique of tigers is to follow the scent of a victim, use hearing and vision to slink stealthily to within about 20–30m (65–100ft), and then charge at speed. The tiger then bowls over the prey, slashes with its huge-clawed forepaws, and sinks its teeth into the prey's neck or back. Most victims are large, sometimes equalling the tiger's own weight, and may include bigger deer, wild cattle, and pigs.

Sadly, if one species is to symbolize the plight of Asian mammals, it is the tiger. Despite many laws put in place to protect it, as well as numerous reserves and sanctuaries, armed guards, and massive local efforts in education and conservation, tigers unfortunately continue to be poached at estimated rates of one every two days.

"To lose the **tiger** would be to destroy our **national symbol** of wildlife and a fatal blow to our **ancient heritage.**"

INDIRA GANDHI, 1917–1984

PROFILE

ASIAN BIG CATS

Asia contains six out of the world's seven big cats: the leopard, everywhere but the north; the tiger, in the south and east; the Asiatic lion, in northwest India; the snow leopard, in the Himalayas; the Eurasian lynx, in the north; and the clouded leopard, in the tropics.

SNOW LEOPARD

EURASIAN LYNX

CLOUDED LEOPARD

POUNCING ON PREY
After perhaps an hour of cautious stalking, a tiger bursts from concealment and covers the final 20-30m (65–100ft) to its prey in just three or four bounds.

169

SLOW AND CAUTIOUS
The slow loris's grip is so powerful it can walk upright or upside down with equal ease along a branch. It moves only one limb at a time, while peering around using its extraordinarily flexible neck.

ENJOYING THE HIGH LIFE

Safe out of reach from the ground-bound predators (tigers, jackals, and dholes), dozens of primate species jump, swing, and chatter through the branches in the tropical forest. Asia has representatives from all three major primate groups: lorises from the prosimians; about 45 species of monkeys; and 17 ape species. In fact, Asia's primate gallery goes further. The southeast islands are homes to some five species of tarsiers (pp.146–47), which outwardly resemble lorises and bushbabies, but have features of both prosimians and monkeys, and are variously classified with either of these main groups. In addition, Asian tropical forests are the only place where lesser apes are found – there are 15 or so gibbon species (see box, above) – as well as two species of greater apes, the Bornean and Sumatran orang-utans.

Lorises have many adaptations to nocturnal hunting. Their huge eyes take up more than half the head and allow acute vision at night. All four limbs have a first digit, or "thumb", that opposes the others, enabling a vice-like grip around small branches. Lorises are slight and move with glacial slowness as they creep towards a likely meal such as an insect, lizard, or baby bird. Then, in contrast to its leisurely stalk, the loris lurches its body forward and throws out its front limbs to grab

In Southeast Asia the dawn chorus is dominated by the whoops and howls of gibbons. Most species form enduring female–male partnerships. Each morning, the pair sit near each other and call loudly in "duet" format, answering each other's song. The chorus proclaims ownership of their territory.

LARGEST AND LOUDEST
The largest of the gibbons, siamangs have a large vocal sac that expands to allow loud and varied cries.

the victim in its hands, while keeping an anchored grip on the branch with its rear feet. It even follows this procedure for prey that is unlikely to run away, such as a tree snail. Lorises also eat soft vegetation such as shoots, buds, and fruits. The two species of slow loris (see top left) are about the size of a domestic cat and their ranges extend from India and Sri Lanka down into the Malay Peninsula and southeast islands. The slender loris (see opposite) is about half the size of the slow loris and is confined to India and Sri Lanka.

MACAQUES AND SURELIS

The most typically Asian monkeys are macaques, with all but one of the 20 or so species found here. (The exception is the barbary macaque or ape, p.44, of North Africa and Gibraltar.) Macaques are strong, sturdy monkeys, and many live part-time on the ground. The most common species is the crab-eating macaque of the Indo-China Peninsula and southeast islands (p.161). It eats a huge variety of foods, from seeds and fruits to fish and crustaceans. It is at home climbing in trees, swimming in creeks, running on the ground, and near human habitation, which has enabled a wider distribution. Its adaptability contrasts with the lion-tailed macaque of monsoon forests in southern India, which is increasing threatened as its habitat is cleared for farmland.

The lowland forest-dwelling sureli monkeys of the Malay Peninsula and large islands, such as Sumatra, Java, and Borneo, are more adapted to life in the trees than macaques. They are slimmer and more lightweight, with long, slender hands equipped with powerful, hook-like fingers. The eight species eat mainly fruits and seeds and have probably all evolved from a common ancestor, geographically separated from the others by water or mountains.

Most primates are social and form close-knit troops. Status is maintained by a variety of methods including calls, threat displays, and mutual grooming. In Asian monkeys, such as langurs or leaf monkeys, about one hour daily is spent in grooming activities. In general, senior members are groomed more, and groom others less, than juniors. Sub-groups or coalitions within the troop also groom each other more often than more solitary members of the group. This intimate activity has the practical benefit of removing dirt, tangled fur, and pests such as lice and fleas.

YOU SCRATCH MY BACK ...
Javan langurs show typical grooming posture as one, probably of higher status, submits to the attentions of a junior. This is one of the rarest langurs, with only 2,000–3,000 remaining.

FALLING FOODS
Monkeys feed wastefully and shower fruits onto the forest floor, benefiting wild pigs such as this babirusa.

ON LEATHERY WINGS

Asia holds two records from the second largest of all mammalian groups – bats. Of the 1,000 species or so occurring worldwide, about one-third live in Asia, including the largest and the smallest. The tiniest is Kitti's hog-nosed bat from Thailand, which was discovered for science as recently as 1974. The head and body are as small as a human thumb-tip, the tip-to-tip wingspan is 15cm (6in), and it weighs only 1.5-2g (less than ⅙oz), making it the smallest of all mammals. One thousand times heavier are Indian fruit bats, which have wingspans exceeding 1.6m (5¼ft). These two species represent the two main groups of bats: the microchiropterans (known as microbats) and the megachiropterans (known as megabats), which differ significantly in diet and way of life as well as in size.

MEGABATS

Most of the 170 or so species of megabats are found in southern and Southeast Asia, with some species in Africa and Australia. They are also known as fruit bats, because many eat fruits and associated plant material such as flowers, sap, buds, shoots, and seeds, according to the season. Some species, mainly those of the genus *Pteropus*, are called flying foxes (see opposite), because of their long-muzzled, vaguely dog- or fox-like faces

FRUITY FOODS
Most fruit bats lap up nectar and pollen, or bite off chunks of fruit. Instead of chewing it, they squeeze the fruit between the tongue and ridged roof of the mouth, swallow the juices, and spit out the squashed pulp.

and often reddish fur. They generally fly and feed from dusk into the night, and roost by day. Unlike microbats, only a few megabat species, mainly the cave-dwelling rousette fruit bats, echolocate to find their way around, and this mechanism is very unsophisticated in rousettes compared to microbats (see opposite).

Many fruit bats roost in large colonies of thousands or, in the case of rousettes, sometimes more than a million individuals, hanging from tree branches. At dusk across Asia, fruit bats of all kinds flap and whirr into the air in vast clouds, to start their foraging trips. As they return to favoured roosts, they jostle and snap at each other to gain the safest positions towards the middle of the hanging mass. Rousette fruit bats consume mainly nectar and pollen, and must fly 40km (25 miles) or more every day in order to gather enough food for survival.

In some areas, especially India and the Malay Peninsula, fruit bats are considered locust-like pests. They descend on a farm field or orchard and within an hour strip it of ripe produce which was ready to pick for market. However, while fruit bats cause crop destruction, they are also a vital part of the forest ecosystem. As they nose into tree blossom and flowers they transfer pollen, and their messy eating habits and droppings spread seeds far and wide.

"At dusk across Asia, fruit bats of all kinds flap and whirr into the air in vast clouds, to start their foraging trips."

THE DARKENING SKY
Fruit bats leave their roosts by the thousand in the early evening. They locate ripe food or flowers in the trees, mainly by using their sense of smell.

Social interactions at fruit bat tree roosts can be extremely complex. In many species, dominant males take the most favoured roosts towards the upper centre of the tree, which are less exposed to attack from predators such as eagles. Each defends a "harem" of up to ten females, and flaps or bites at attempted takeovers from other males, who usually cluster in bachelor groups.

INDIAN FRUIT BATS
One of the largest megabats, this species may use the same trees year-round, for 30 years or more.

The smallest megabat is the pygmy fruit bat of the Malay Peninsula and nearby islands such as Sumatra, Java, and Borneo. Its head and body are just 7cm (3in) long and it weighs only 20g (¾oz). These tiny megabats frequent upland forests at 1,000-2,000m (3,300-6,550ft) altitude, and unusually for fruit bats, often roost alone.

MICROBATS

In contrast to fruit bats, the Asian microbats are generally insect-eaters. Like their larger cousins, they have forelimbs adapted as wings, flapped by powerful pectoral chest muscles. Unlike most megabats, the microbats can emit pulses of ultrasound through the mouth and/or nose, and then hear the echoes bounced back by nearby objects. Echolocation helps the bats to twist and turn among forest twigs and leaves, and pursue prey such as moths and gnats, all in complete darkness. Especially powerful and agile fliers are Asian false vampire bats. As their name suggests, they do not suck blood like the true vampire bats of North, Central, and South America (pp.100–101). But they are efficient hunters and snatch large insects such as crickets and cockroaches, as well as mice, fish, and frogs from forest pools, small birds from their roosts, and even smaller bats in mid-air. They are even known to fly through open windows or doors into rooms, pick a gecko or other lizard off the wall, and leave again – all within a second.

BLACK FLYING FOX
This typical fruit bat wraps its wings around its body while roosting – except in very hot weather, when it opens and flaps its wings to create a cooling breeze.

EUROPE

Iceland

400–200 MILLION YEARS AGO
Much of Europe is at the equator,
joined to North America in the west
and to Asia in the north. At this
stage, Africa butts up to the Iberian
Peninsula. Europe forms part of the
supercontinent Pangaea.

50 MILLION YEARS AGO The
Atlantic Ocean begins to form
as Europe and North America
drift apart. A seaway separates the
eastern edge of Europe from Asia.

30 MILLION YEARS AGO The
northward drift of Africa begins
to form the main mountains
of southern Europe.

2 MILLION YEARS AGO The last
series of ice ages begins to sculpt the
Alps as we know them today. Sea
levels fall, exposing more islands
in the Mediterranean Sea.

100,000 YEARS AGO Ice covers
more than half of Europe, as far
south as what is now England
and across to northern Germany
and the Balkans. Glaciers spread
from the Alps and other mountains.

40,000 YEARS AGO Central Europe
is populated by ice-age mammals,
such as woolly mammoths, woolly
rhinos, cave bears, Irish elk, and
Neanderthal people, who have
been around for 200,000 years
but will soon die out. Modern
humans (Cro-Magnons), originally
from Africa, spread from the
eastern Mediterranean into
southwest Europe.

**PREVIOUS PAGE:
BROWN HARE**
When Europeans colonized lands,
they took their animals with them.
The brown hare is now found on
every continent except Antarctica.

HABITATS OF
EUROPE

EUROPE IS A RELATIVELY SMALL CONTINENT, SURROUNDED
ON THREE SIDES BY WATER. IT SHARES A WIDE BORDER WITH
ASIA TO THE EAST, SEPARATED BY MOUNTAINS AND SEA, AND
HAS CLOSE LINKS WITH NORTH AFRICA TO THE SOUTHWEST.
THE CLIMATE IS GENERALLY MODERATE, SO IT LACKS
EXTREME HABITATS SUCH AS TRUE DESERT.

A TEMPERATE HABITAT
Politically, Europe's boundaries have moved many
times in history, and continue to change. But its
natural boundaries remain watery on three sides –
Barents Sea to the north, the Atlantic Ocean to the
west, and the Mediterranean Sea to the south. To the
east, Asia is separated from Europe by the Black Sea,
the Caucasus and Ural Mountains, and the Caspian Sea.

Most of inland Europe is temperate: it has a
mild climate with moderate rainfall and mid-range
temperatures; it is cooler in the north than it is in
the south. This encourages the growth of a great
variety of vegetation, from conifer forest in the north,
mixed deciduous woodland in central areas, and shrub,
scrub, and patchy grassland in the hotter, drier south.
Iceland, the Mediterranean islands, and northwest
North Africa all share plant and animal species with
Europe, and can be regarded as biogeographic
extensions of the continent.

HUMAN IMPACT
Europe has a very long history of human occupation –
few other places on Earth have been dominated for
so long by so many people. As a result, human impact
is very apparent. Most of the land has been turned
over to farming or to the creation of quarries, factories,
housing, and modern leisure facilities, and very few
sizeable areas of true wilderness remain. This long
history of human domination, combined with its
diminutive size (excluding
Australia it is the smallest
continent) and moderate
climate, has contributed to
the relatively small number
of unique or notable
mammal species that
Europe supports. Hunting
and direct persecution have
had an effect over the
centuries; but habitat loss
is probably the most
significant factor, and larger
wild mammals in particular
face an uncertain future.

KEY TO HABITATS

- GRASSLAND
- MOUNTAINS
- WETLANDS
- FOREST

Faeroe Islands

Outer Hebrides

Ireland

British Isles

ATL

Brit

Shannon

Tha

English

Loire

Bay of
Biscay

M
C

Garonne

Pyrene

Douro

Iberian

Ebro

Tagus

Peninsula

ALP

Guadalana

Ba
Is

Guadalquivir

ATLANTIC OCEAN

M

Medi

GRASSLAND (PP.178–79)
Mosaics of fields, meadows,
pastures, and cereal croplands
are dotted across much of
Europe's middle zone, where the
climate is most equable – winter
snow lasts for a few weeks at
the most, and summer droughts
are equally short.

WETLANDS (PP.182–83)
Wetlands in Europe vary from cold lakes and marshes in the north, which ice over for months in winter, to warm swamps and lagoons in the south, which dry out during the hot summer into cracked mud.

FOREST (PP.184–89)
Without human interference much of Europe would be woodland. Of what remains, the natural vegetation varies from dense forests of needle-leaved evergreens in the north, to open glades of broadleaved trees in central areas, and scattered groves of drought-adapted shrubs fringing the Mediterranean.

MOUNTAINS (PP.180–81)
A band of mountains and uplands runs across much of southern Europe, from the Cantabrians and Pyrenees in northern Iberia, east through the Alps to the Carpathians and Balkans. Mammals adapted to high altitude are joined in the mountains by lowland species driven to more remote areas by habitat loss and human persecution.

ESSENTIAL INFORMATION

■ **SIZE** 10.5 million square km (4 million square miles), including the European portion of Russia, making up 7 per cent of Earth's land area.

■ **NATIONS** About 40 independent states, including some of the world's smallest (Liechtenstein, Andorra, and Vatican City).

■ **CLIMATE** Fairly moderate, but ranging from high Arctic in northern Scandinavia (average annual temperature –3°C/26°F) to

dry Mediterranean (average summer temperature 28°C/83°F).

■ **HIGHEST MOUNTAINS** Mt Elbrus (Caucasus, bordering Europe and Asia), 5,642m (18,510ft) high; Mt Blanc (Alps), 4,807m (15,770ft) high.

■ **LONGEST RIVERS** Volga (into Caspian Sea, bordering Europe and Asia), 3,690km (2,290 miles) long; Danube (into Black Sea), 2,850km (1,770 miles) long.

GRASSLAND TYPES
In the far south, especially along the Mediterranean coast, are dry, scrubby, chaparral-type landscapes, where flowers and herbs bloom briefly before the long, hot summer. The natural grasslands of northern Europe include damp hay meadows, where pools of water ice over in winter. However, a mixture of arable and livestock-grazed farmland is now dominant in Europe.

GRASSLAND
COMPARED TO THE VAST SAVANNA, PRAIRIE, AND PAMPAS OF OTHER CONTINENTS, EUROPE'S GRASSLANDS ARE GENERALLY RELATIVELY SMALL AND PATCHY, SET IN WOODLAND OR FOREST. HOWEVER, IN THE SOUTHEASTERN FRINGES, TREELESS FLATLANDS MERGE INTO THE GREAT STEPPES OF ASIA.

MEADOW AND PASTURE
Europe's natural grasslands are few and far between, mainly because the climate has enough annual rainfall, without a long summer drought, for trees to thrive. Apart from smaller areas on hillsides and uplands, the main natural regions of lowland grass are the temperate steppes north of the Black and Caspian Seas, and the tough, scrubby, chaparral-type dry grasslands along the Mediterranean coast. Consequently, European wild mammals include few large, open-habitat grazers, and most grassland

mammals are small and relatively inconspicuous. However, bigger browsers and rooters, such as deer and wild boar (p.184), may wander from woods into grassy clearings, to nibble at the young shoots and leafy blades.

Many deer are woodland browsers but some species, for example fallow and red deer, tend to use the trees mainly for shelter, and graze on grasses and rushes in adjacent open areas, such as meadows, pastures, arable fields, and parkland. These deer feed mainly around twilight, resting during the middle

HARE IN MID-AIR
A frightened brown hare can easily cover 3m (10ft) in one leap, and then on landing can jump away at right angles, to throw off a pursuer.

of the night and day. In areas where human disturbance is rare, they may lie in the open to rest by day. If they sense danger, they will leap up and bound into a thicket or tree clump. In the rutting season of autumn, fallow bucks gather at traditional sites called stands, or leks, in field corners; on meeting their rivals they snort, clash antlers, and stamp their hooves.

Most of Europe's original low-lying grassland, in common with much greater areas of forest, have long been under the plough. The patchwork of meadows and pastures for grazing livestock, such as cows and sheep, interspersed with wide fields of cereal crops, like wheat and barley, and separated by hedges and low stone walls, has developed over many centuries. Newer forms of planted grassland include parks and gardens, road verges, and leisure areas such as golf courses.

The farming landscape is now so established and familiar that many smaller mammals have adapted to it. There are plentiful rodents, such as rats, mice, and voles, including one of the smallest mammals, the harvest mouse, which is the only European rodent with a prehensile tail. This species builds its tennis-ball-shaped nest among long stalks in a grassy clump. Most of the other rodents dig burrows under grass roots, where they hide by day and emerge under cover of darkness to feed.

OPEN-COUNTRY LEAPERS

Outwardly similar to rodents, but members of a separate mammal group known as lagomorphs ("leaping forms"), are hares and rabbits. The brown hare in particular (see opposite and pp.174–75) is highly suited to open habitats. It is able to race from danger at over 70km (44 miles) per hour,

with powerful thrusts of its long, muscular rear legs. Unlike its cousin the European rabbit (see right), which prefers family life in a network of tunnels called a warren, the hare has a solitary lifestyle. It nibbles grasses and herbs at dawn and dusk, and it rests by day lying flat among the grasses in a shallow depression known as a form.

Two predators, in particular, live off the rodents and lagomorphs, namely the weasel and the stoat (see below). Both belong to the mustelid group of carnivores. They are short-legged, long-bodied, and sinuous, and they hunt in similar ways. The two avoid competition by taking different sizes of prey. The weasel's main prey is smaller herbivores, such as mice, voles, baby rabbits, and young birds. It has a head-body length of about 20cm (8in), and a very slim head with pointed muzzle. It is so slender that it can enter a mouse burrow with ease. The stoat is a few centimetres longer and more bulky and muscular and, unlike the weasel, it has a black-tipped tail. Stoats tend to take larger victims, such as rats, young hares, and adult rabbits, as well as medium-sized birds such as partridge.

CARNIVOROUS STOAT
Active day and night, stoats may cover 10km (6 miles) on a single hunting foray. They can rear up on their hind legs to peer over taller vegetation.

THE SMALLEST HUNTERS

The predators of grassy southern Europe also include the smallest of all terrestrial mammals, the pygmy white-toothed shrew, also known as the Etruscan shrew or Savi's shrew. Its head and body are as small as a human thumb, and it weighs just 3g (⅒oz). Its fractionally larger cousin, the Eurasian shrew, is widespread across most of the continent. These tiny, quivering bundles of ferocious energy must eat almost their entire body weight every 24 hours or starve. Their prey includes grubs, worms, slugs, snails, spiders, and insects up to the size of grasshoppers.

FAMILY GROUP
Originally from grasslands in Iberia, European rabbits live in extended family groups in communal burrows (warrens), with many entrance holes.

PIPISTRELLE BAT
The pipistrelle is Europe's smallest bat. It is probably also the most common. It flies out from churches, barns, and other quiet buildings at dusk, to hunt for moths, midges, and gnats, which flit above the grasses.

LIFE UNDERGROUND   BEHAVIOUR

In grassy habitats, the rich soil teems with root-munching grubs, bugs, millipedes, slugs, and worms, which are all food for the European mole. It digs a tunnel system using its large-clawed, shovel-like front paws, and patrols the tunnels regularly for small creatures. The main nest is in a larger chamber near the centre of the system, which may have a total tunnel length of 500m (1,650ft).

MOUNTAINOUS MOLEHILLS
A tunnelling mole pushes up excess earth as molehills, which dot many meadows and pastures, and even well-manicured lawns.

MOUNTAIN TYPES
On the higher mountains of Europe, winter avalanches and summer landslides of cracked, loose rocks, called scree, regularly sweep away vegetation. On the lower slopes, less extreme climates allow grasses and shrubs such as heather to flourish.

MOUNTAINS EUROPE'S PEAKS

CANNOT COMPARE WITH MAJOR GLOBAL RANGES SUCH AS THE HIMALAYAS, ANDES, AND ROCKIES. BUT THEY ARE HIGH ENOUGH AND SUFFICIENTLY EXTENSIVE TO SUPPORT A GREAT NUMBER OF THICK-FURRED, NIMBLE-FOOTED, CRAG-LEAPING MAMMAL SPECIES.

ALPINE HERBIVORES

Europe's chief mountain range, the Alps, has been thrust up relatively recently and is still growing, at about 1cm (½in) yearly. Its young peaks are sharp and jagged, not yet rounded down by the erosion of ice, wind, sun, and rain. Slopes above about 2,500–3,000m (8,200–9,850ft) are covered in snow much of the year and are almost bare of life. Medium altitudes of 1,500–2,500m (4,900–8,200ft) are a mixture of rocky outcrops, plateau meadows, and grassy inclines. Wild goats, such as the chamois (see box, below) and ibex (see opposite), and the wild sheep known as the mouflon (see opposite), as well as numerous domesticated goats, sheep, and cattle, trek to these upland pastures during spring. They feast on the herbs, grasses, and low shrubs, then descend in autumn through the conifer tree zone, which reaches about 1,600m (5,250ft) high, to the shelter of the mixed deciduous forests that cloak slopes up to around 1,300m (4,250ft) high.

HIGH-ALTITUDE GRAZERS

Smaller herbivores, including a type of ground squirrel, the alpine marmot (see above), frequent the high meadows, mainly in the Swiss–Italian Alps. These marmots avoid the dangers of a long trek down the mountain in autumn by hibernating instead. During their six-month winter in a communal den, where they live in extended family groups of 5 to 15, marmots lose one-third of their body weight. Through the summer months they risk attack by foxes, wolves, pine martens, and golden eagles, as they feed by day on grasses, herbs, and sedges, along with seeds and roots in autumn. The mountain hare (see left), found in isolated pockets up to 1,300m (4,250ft) high in the Alps, and also across most of upland northern Europe, is another summer grass- and herb-grazer and

ALPINE MARMOTS
Once hunted for their body fat, alpine marmots have been re-established in the Pyrenees and Carpathians. In battles over territory they rear up, growl, screech, and scratch.

turns to bark and twigs of woody shrubs in winter. Its alternative names of variable hare and white hare signify its autumn moult from brown to almost white coat, for winter camouflage in ice and snow.

MOUNTAIN PREDATORS

There are three large mammalian predators in Europe: the brown bear (p.185); the grey wolf (p.90); and the Eurasian lynx (see left and p.169). They are still relatively common in the far north and east, but are very rare in the more heavily populated southern and western areas, where they have been persecuted for attacks on farm animals and even people. Those that remain have found refuge in the mountains. Small populations of brown bears still exist in northern Iberia towards the Pyrenees, the Apennines of central Italy, and the Dinarics and Balkans to the east. Grey wolves have a similar, very limited distribution. The Eurasian lynx, although widespread in the northeast, was exterminated in the Alps by the early 20th century. In the 1970s it began to be reintroduced in Switzerland, Slovenia, and the Jura Mountains.

For such reintroductions to be acceptable to all, livestock must be protected. This might necessitate farmers in southern and western Europe returning to the old stock-guarding traditions that are still followed in the north and east. This would make it possible for predators and farmers to co-exist.

EURASIAN LYNX
After being reintroduced to the central Alps, lynx now number more than 100. They hunt roe deer (p.184) and also chamois (see box, right), which are themselves scarce.

MOUNTAIN HARE
To reduce the risk of frostbite, the mountain hare has shorter ears and limbs than its lowland cousin (p.179).

MOUNTAIN DWELLERS

CATS, such as the wildcat, hunt for rodents, rabbits, birds, and reptiles, in rocky clearings and woodlands on lower slopes.

MOUNTAIN MUSTELIDS, such as the wolverine, are found in colder uplands of northern Europe, usually in conifer forests.

HOOFED MAMMALS, such as the endangered chamois, inhabit high to mid-altitudes in summer and mid- to low in winter.

Today's domestic breeds of sheep descend from the Asiatic mouflon of southwest Asia (p.153). Ancient people brought mouflon herds to the Mediterranean islands, then into southern and central Europe, leaving a trail of part-domesticated sheep that then returned to the wild. Soay sheep of the remote St Kilda islands, off northwest Britain, resemble those wild ancestors.

ORIGINAL SHEEP
Viking travellers may have taken sheep to St Kilda over 1,000 years ago. They now live wild and retain their ancestral small size and dark brown coats.

"Many large mammalian predators have found refuge in the mountains: the brown bear, the grey wolf, and the Eurasian lynx."

RETURN TO THE PEAKS
In the 19th century, alpine ibex were reduced by hunting to just one herd. However, as a result of a reintroduction programme they now number tens of thousands along the Alps chain.

FEEDING TIME
Water voles prefer greenery such as shoots and young leaves but will also gnaw soft bark and buds. Occasionally, they will feed on small fish such as minnows.

WETLANDS
MORE THAN TWO-THIRDS OF EUROPE HAS MODERATE RAINFALL AND LOW, WELL-ROUNDED TOPOGRAPHY. THE CONTINENT'S MAJOR RIVERS ARE WIDE AND TEND TO RUN SLOWLY MOST OF THE YEAR, AND THE MAJORITY OF MARSHES AND SWAMPS RARELY FLOOD EXCESSIVELY OR DRY OUT.

WATERY HABITATS
Europe's dampest lowland regions are the west coasts of Scandinavia, the British Isles, northern Iberia, the Italian Peninsula, and the Balkans, where the landscape is lush and occasionally marshy or boggy. Rainfall decreases to 500–1,000mm (20–40in) across vast areas of the continent's centre, which is rarely enough to form extensive wetlands, although in the cooler north, where evaporation is low, great lakes dot the landscape around the Baltic Sea. In the hotter south, wetland habitats are confined to the main rivers and a few swampy deltas near the coast, such as the Coto Doñana on the Guadalquivir River in southwest Spain, the Camargue at the Rhone's mouth in southern France, and the Dobruja marshlands, where the Danube enters the Black Sea.

WATER–DWELLING MAMMALS
Even before people settled around Europe's fresh waters there were few large aquatic mammals. However, rodent and insectivore species abound, where permitted in the face of increasing pollution and waterway development for irrigation, human consumption, and watersports.

On a calm evening, a V-shaped ripple might signify a brown rat, a water vole – also known as a water rat (see opposite), or perhaps a water shrew (see above). The water vole has a short muzzle, chubby body, thick brown coat that almost hides its small ears, and a slightly furred tail that is shorter than that of the brown rat. Water voles space themselves in territories at 100–200m (330–660ft) intervals along a bank. With their generally slow and placid lifestyle, they may fall prey to the sudden strike of otters (see right), mink, large pike, owls, or large wading birds such as herons.

The water shrew, like the water vole, lives a mainly solitary life based in its small burrow system dug into the bank, with usually one or two openings under the water. It also paddles with all four legs and rarely stays under the surface for more than 30 seconds. But in contrast to the mainly herbivorous vole, the water shrew is a fast-moving hunter with a pointed, quivering muzzle

that searches out small creatures, such as pond snails, mud-dwelling worms, aquatic insects, small fish, tadpoles, young frogs, and even waterbird chicks.

WATER SHREW
One of the few venomous mammals, the water shrew has saliva containing a mild toxin that subdues prey. Its rear feet are fringed with bristly hairs.

BANKSIDE HUNTERS
Chief among the mammal predators of European wetlands are the mustelid mammals, such as otters and mink. The Eurasian otter (see below right), which disappeared totally from many areas in the 20th century as a result of pollution, over-defensive anglers, fur-trappers, and sport-killing, is slowly returning to its previous strongholds. Otters screech in competition for individual territories of about 5km (3 miles) of bankside, extending to 20km (over 12 miles) where prey density is low. The otter's webbed paws, waterproof fur, and stiff whiskers that enable it to gauge currents and sense prey in the water, equip it superbly for aquatic hunting.

Mink are considerably smaller than otters, with less waterproof fur and partially webbed feet. From about 1900, American mink (p.76) escaped from European fur farms and spread, being adaptable hunters in water and also on land. The slightly smaller European mink has suffered from habitat interference by humans and competition from its American cousin. As a result, the European mink is one of the continent's most threatened mammals.

WETLAND TYPES
Plentiful rain produces fast-running streams on hilly slopes, mainly in the far west of Europe. More extensive are the lowland ponds and lakes, especially in the colder north.

EURASIAN OTTER
In winter, the Eurasian otter forages up to 10km (6 miles), usually after dusk or around dawn. When ice cuts off its supply of fish and frogs, it eats rodents and small birds.

SKIMMING AT TWILIGHT
BEHAVIOUR

Wetlands are "insect nurseries", with larvae of mayflies, caddisflies, and many other species emerging from the water to become winged adults. Several bats, for example Daubenton's, also known as the water bat, specialize in skimming the surface of a variety of wetland habitats for flies, moths, and other airborne prey. These are caught in the mouth or are scooped into the curled, pocket-like tail membrane. Daubenton's bat is also adapted for catching fish. It detects water ripples with its echolocating sonar, and uses its large back feet and its claws to grab fish swimming just below the surface of the water.

ON THE WING
By day, Daubenton's bat roosts in trees and buildings, but at twilight it heads off to feed. In winter, it flies up to 300km (180 miles) to hibernate in a cave or mine.

BROADLEAVED WOOD

CONIFER FOREST

FOREST TYPES
Europe's main central band of temperate broadleaved woodland includes beech, oak, birch, elm, and chestnut trees. The northern conifer forests of needle-leaved evergreens consist mainly of firs, pines, and spruces.

FOREST
EUROPE HAS THREE MAIN WOODED ZONES: VAST CONIFER FORESTS, ACROSS THE NORTHEASTERN QUARTER; SUBTROPICAL SHRUBBY EVERGREENS, ALONG THE SOUTHERN COASTS; AND MIXED DECIDUOUS WOODLANDS OF BROADLEAVED TREES, WHICH GROW ALMOST EVERYWHERE ELSE ON THE CONTINENT.

NORTHERN AND EASTERN FORESTS
Before much of Europe's land was ploughed for crops and grazed by livestock, trees – from small clumps to vast forests – were the predominant natural vegetation. Remnants of those once-great wild lands are preserved in isolated areas, but few extensive tracts persist today. The most substantial areas of forest exist mainly in the remote north and east of the continent, where the larger wild mammal species dwell.

The largest European wild mammal inexistence is the European bison, or wisent (see left). Other large mammals are the elk (or moose as it is known in North America, pp.80, 88–89), deer, and wild boar (see above). The chief predators of these large herbivores are grey wolves (p.90), although brown bears (see opposite) may occasionally turn from their largely vegetarian diet and attack these species. Recent genetic studies suggest the European bison may be the same species as the American bison (pp.68–69). It has a slightly lighter, shorter

WILD BOAR
A sounder (group) of wild boar usually consists of sows with their last two years of boarlets. Males tend to be solitary (especially older boar), except during the mating season in mid-winter.

coat than that of the American bison, but it is otherwise very similar. Hunted to extinction in the wild by about 1910, European bison were park-bred and reintroduced into the mainly coniferous Bialowieza Forest, on the Poland-Belarus border, and also into reserves in the Caucasus Mountains. Large males weigh close to 1 tonne (2,200lb) and live alone or in bachelor groups, except during the rutting season (between August and October) when they engage in a violent conflict of clashing heads and horns: the winners of these contests then join the females, which live with their most recent offspring in small herds of between 5 and 15, with one dominant elder (the matriarch). After mating, the groups segregate again into bachelors and mothers with calves.

EUROPE'S BEARS
In southern Europe, bears are extremely scarce, historically banished by human persecution to remote rocky uplands (p.180). They become progressively more common to the north and east, where natural forest habitats increase and human populations decrease. These Eurasian bears are subspecies of the brown bear; they are also the smallest – males rarely exceed 250kg (550lb), and females 150kg (330lb). The brown bears of Alaska (p.81) and northeast Asia may be four times heavier than the bears found in Europe.

Averaged through the year, some 90 per cent of the brown bear's diet is vegetarian. It switches through the seasons to consume roots, bulbs, tubers, shoots, buds, grasses, leaves, fruits, berries, and nuts – almost any plant matter. When it feeds on meat it takes either very small animals, such as grubs from rotting logs or larvae in ants' nests, or much larger, often dead animals, such as scavenged deer carcasses. In the colder parts of its range, brown bears may become dormant in winter, retreating to their dens and living off their body fat. It is usually at this time that the female gives birth to her young.

The brown bear is short-sighted but its hearing and sense of smell are very keen. Despite its size it is quite agile, and is especially renowned for its tendency to stand upright when threatened. It often ambles along at around 5km (3 miles) per hour, but if disturbed it may suddenly charge at ten times this speed.

BISON FEEDING
European bison, or wisent, gorge on acorns, beech mast, and seeds in autumn and eat evergreen shrubs, mosses, and lichen in winter.

FOREST SIGNPOSTS
BEHAVIOUR

Woods and forests are full of signs left by mammals marking their territory. Smaller deer, such as the roe, rub fluid from scent glands on the head and between the hooves onto trees, bushes, logs, and stones. Male deer also thrash or "fray" their antlers into a bush or shrub, partly to get rid of old skin on their antlers but also to leave broken twigs and leaves as an indicator of their presence

MARKING THE TERRITORY
A roe stag marks his territorial boundary with pre-orbital scent glands situated just under the eyes.

A LONG WAY DOWN
Adult brown bears (below) are mostly too weighty to climb trees. However, cubs (left) clamber up in their urge to explore. Most cubs are weaned by 18 months but stay with the mother for another year or so. The pale "collar" of fur persists until sexual maturity (5 to 7 years old).

FLYING LEAP
Red squirrels can jump 2m (6½ft) or more as they leap from one branch to another. If they lose their footing for any reason they can survive falls from considerable heights.

OUT INTO THE WILD WOOD
Red foxes live as a vixen-dog pair in their den called an earth, and both parents care for their four to five cubs. The young are brown at first but gain their reddish coat by 8 weeks, when they start their early hunting forays.

THE WOODLAND TRAIL

Most of western Europe's pockets of broadleaved woodlands survive today because they have been actively conserved for specific reasons, mainly for the benefit of humans, rather than being left alone by accident or neglect. Hunting for large mammals, such as deer and wild boar, has long been a prime reason. Other modern woodland uses tend to be a mixture of habitat and wildlife conservation, managed hardwood production, and leisure pursuits. Walkers on woodland trails can hope to glimpse mammals such as badgers, red foxes (see opposite), deer (p.184), stoats (p.179), hedgehogs, and varied smaller species of shrews, voles, and mice.

NATIVES AND IMPORTS

Among the mammals most suitably adapted to tree living are the sciurids, or squirrels, which live on seeds, nuts, flowers, and fungi, as well as tree bark and sap. Europe has two species: the red squirrel (pp.186–87) and the grey squirrel (see right and p.83). The red squirrel's fur, which is reddish brown in summer, moults to deeper brown tinged with grey in winter. Apart from colour, the red squirrel can be readily distinguished from the grey by its smaller size, bushier tail, and tufted ears.

The grey squirrel was introduced to various sites in Britain from North America in the 1870s, and in many areas it now thrives at the expense of its red cousin: grey squirrels in Britain now number more than 2.6 million while there are fewer than 150,000 reds. The greys have also gained a foothold in small areas on the European mainland, especially in Italy, although numbers of red squirrels there are still fairly high.

The grey squirrel is not the only reason for the red squirrel's decline. Even where greys are absent, numbers of red squirrels have reduced in some areas, as a result of virus-based disease and loss of their more specialized habitats, especially hazel woods and native conifers.

COMMON DORMOUSE
Also known as the hazel dormouse, this mouse climbs nimbly, using its furry tail to balance, and rarely descends to the ground. Its diet includes seeds, berries, nuts, fruits, and grubs.

Europe is host to various other mammal introductions, including the sika (p.150), chital or axis deer (pp.150–51), and muntjac deer from Asia, and the white-tailed deer (p.80) from North America. Moving in the opposite direction, the red fox has been introduced by Europeans to other regions such as Australia (p.131). This is one of the world's most widespread and adaptable carnivores. In Europe, it is regarded as useful by some people since it keeps down numbers of rabbits, mice, and other herbivorous pests. Other people view it as a pest because it raids chicken-houses and spreads the potentially fatal disease rabies.

WOODLAND EDGES

The edges of a woodland, where trees give way to bushes and grasses, provide a mixed, mammal-rich habitat. Hedgerows furnish valuable shelter and linking corridors between woods and copses, for small mammals such as shrews, hedgehogs, mice, and voles, and their predators such as weasels and stoats. The wood mouse, also called the long-tailed field mouse, and the yellow-necked mouse, are widespread across all European deciduous woods and are staple food for mustelids, foxes, and hunting birds, such as hawks and owls. There are another 25 or so species of mice, voles, lemmings, and other small rodents found among European trees. The wood lemming is common in conifer forests in Scandinavia, the bank vole is widespread in deciduous woods throughout the continent, while the grey hamster lives in the small and patchy woods of the far southeast.

There are also five European species of dormouse, ranging from the fat or edible dormouse, which may reach 300g (10oz) as it feasts on nuts, berries, and fruits in autumn, to the common or hazel dormouse (see above), which weighs only one-tenth as much. These dormice pass the colder months, generally October to April, in hibernation (see box, left). In spring, the common dormouse weaves a ball-shaped breeding nest, about 15cm (6in) across, of stems and creepers, often founded on an old bird's nest or squirrel drey, a few metres or more up in the branches of a tree. Its larger cousin the garden dormouse, about 100g (4oz), makes a similar structure but lower down, in a shrub, among creepers, in a burrow, or in a roof space or outbuilding.

GREY SQUIRREL
This grey squirrel takes a break from feeding on a hot day to nap in the fork of a tree branch.

BATS

Woodland bats may leave their tree holes before dusk and feed until after dawn. The noctule catches insects while on the wing. Brown long-eared and lesser horseshoe bats grab prey from trees and shrubs rather than catching it in mid-air.

NOCTULE BAT

BROWN LONG-EARED BAT

LESSER HORSESHOE BAT

HIBERNATING MAMMALS

BEHAVIOUR

Many mammals are said to hibernate, but some, such as bears and squirrels, simply sleep more deeply than usual in winter, and rouse during warm spells. During hibernation, body temperature plummets to 3–6°C (38–42°F), and breathing and heartbeat are almost imperceptible. Many rodents hibernate, especially dormice and some marmots (p.180). A dormouse may lose half its body weight in the 6-month hibernation.

HIBERNATING DORMOUSE
Dormice make a nest of stems, twigs, leaves, and moss, among roots or in a burrow.

750 MILLION YEARS AGO Most of the water in today's seas and oceans – some 1.3 billion cubic kilometres (312 million cubic miles) – has formed. Land masses are gathered loosely into an early supercontinent, Rodinia, surrounded by primeval oceans including the Mozambique, the Pan-African, and the Brasiliano.

500 MILLION YEARS AGO Microscopic life has existed in the sea for more than 2,500 million years; now it begins to evolve into complex shelled creatures such as trilobites.

220 MILLION YEARS AGO The break-up of the U-shaped supercontinent of Pangaea begins. This partly encloses the Tethys Sea; Pangaea is itself surrounded by the immense super-ocean, Panthalassa.

130 MILLION YEARS AGO The Atlantic begins to form as a giant inland sea between Europe and North America. There is a seaway between South America and Africa.

50 MILLION YEARS AGO The Red Sea starts to form. Mammals, hitherto restricted to land, now begin to return to the sea as early whales.

25 MILLION YEARS AGO The Pacific Ocean takes shape from the remnants of Panthalassa; the Indian Ocean forms from the leftover Tethys Sea. Early seals evolve.

1.7 MILLION YEARS AGO Sea levels fall as the most recent series of ice ages begin. Levels drop to 130m (425ft) below today's, exposing land bridges that close off many straits. For example, the north Pacific and Arctic oceans become separated at the Bering Bridge.

PREVIOUS PAGE:
SEAL
A common seal saves energy as it dozes on land. Swift and graceful in the sea, out of the water it is slow and clumsy.

HABITATS OF
OCEANS AND SEAS

EVEN THE LARGEST HABITATS ON LAND ARE DWARFED BY THE MARINE BIOME, A VAST EXPANSE OF WATER THAT COVERS MORE THAN TWO-THIRDS OF THE EARTH'S SURFACE. WE MAY THINK OF SEA AND OCEAN CONDITIONS AS INVARIABLE AND CONSTANT, BUT THEY ARE IN FACT FAIRLY ERRATIC.

WATERY WORLD
Life in the mysterious underwater realm can be just as complex and varied as it is in terrestrial habitats. In terms of the number of animal species they support, coral reefs are some of the richest of all habitats. However, our knowledge of marine life is sparse compared to that of creatures and plants on land. Scientific submarine surveys and studies are increasing in number and sophistication, but there are immense regions that remain uncharted. There are doubtless still surprises in store. One such was the discovery, in the late 1970s, of deep-sea hydrothermal vents. These jets of superheated, mineral-rich water spurt from cracks in the seabed and provide nutrition for previously unknown, clustered communities of bizarre worms, fish, crabs, and other creatures.

MARINE MAMMALS
Despite the oceans' vastness and variety, mammals are under represented because of their land-dwelling origins. But through evolution, three mammalian orders (main groups) forsook their land-lubbing past and returned to the sea. These are the cetaceans – whales, dolphins, and porpoises (83 species); the pinnipeds – seals, sea lions, and the walrus (34 species); and the sirenians or sea cows – dugongs and manatees (four species).

These sea-goers exhibit some of the most extreme adaptations from the typical mammalian four-legs-and-a-tail ancestral body plan. Evolution has equipped them with streamlined shapes and paddle-like limbs for swift and efficient movement through the water; fish-like tails or flukes in the cetaceans and sirenians; and a host of specialized features in their body chemistry to enable them to cope with salty water – from drinking it to diving in it.

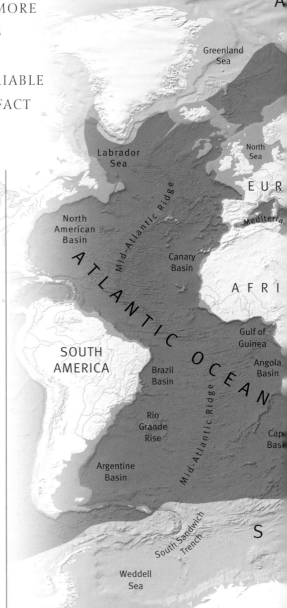

COASTAL WATERS (PP.194–97)
About one-thirteenth of the marine habitat is continental shelf, shallower than 200 metres, where sea life is richest. The shore itself is hazardous, with boulders, waves, and lurking predators

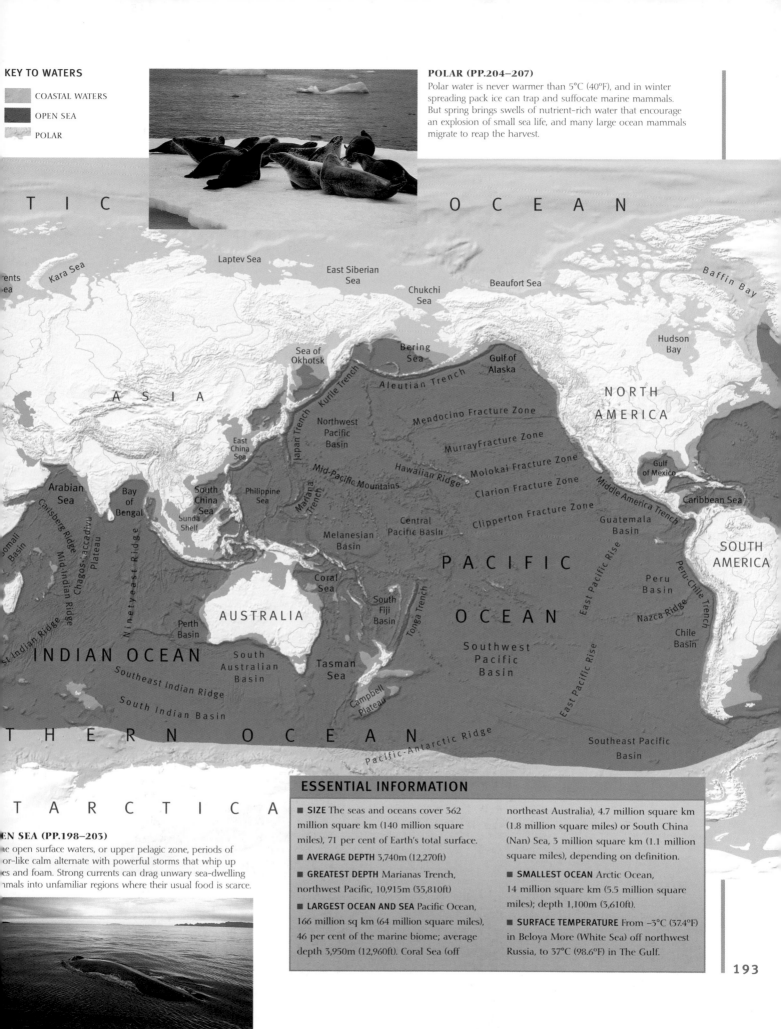

KEY TO WATERS

◻ COASTAL WATERS

◼ OPEN SEA

◻ POLAR

POLAR (PP.204–207)

Polar water is never warmer than 5°C (40°F), and in winter spreading pack ice can trap and suffocate marine mammals. But spring brings swells of nutrient-rich water that encourage an explosion of small sea life, and many large ocean mammals migrate to reap the harvest.

Map labels:

ARCTIC OCEAN

Kara Sea · Laptev Sea · East Siberian Sea · Chukchi Sea · Beaufort Sea · Baffin Bay

rents ea

ASIA · Sea of Okhotsk · Bering Sea · Gulf of Alaska · Hudson Bay · NORTH AMERICA

Aleutian Trench · Mendocino Fracture Zone · Murray Fracture Zone

Kurile Trench · Northwest Pacific Basin · Japan Trench · East China Sea

Mid-Pacific Mountains · Hawaiian Ridge · Molokai Fracture Zone · Clarion Fracture Zone · Gulf of Mexico

Arabian Sea · Bay of Bengal · South China Sea · Philippine Sea · Mariana Trench · Central Pacific Basin · Clipperton Fracture Zone · Guatemala Basin · Caribbean Sea

Carlsberg Ridge · Mid-Indian Ridge · Chagos-Laccadive Plateau · Ninetyeast Ridge · Sunda Shelf · Melanesian Basin · PACIFIC OCEAN · Middle America Trench · SOUTH AMERICA

Somali Basin · West Indian Ridge · Perth Basin · AUSTRALIA · Coral Sea · South Fiji Basin · Tonga Trench · Peru Basin · Nazca Ridge · Peru-Chile Trench · Chile Basin

INDIAN OCEAN · South Australian Basin · Tasman Sea · Southwest Pacific Basin · East Pacific Rise

Southeast Indian Ridge · South Indian Basin · Campbell Plateau · East Pacific Rise · Southeast Pacific Basin

SOUTHERN OCEAN · Pacific-Antarctic Ridge

ANTARCTICA

OPEN SEA (PP.198–203)

...e open surface waters, or upper pelagic zone, periods of ...or-like calm alternate with powerful storms that whip up ...es and foam. Strong currents can drag unwary sea-dwelling ...mals into unfamiliar regions where their usual food is scarce.

ESSENTIAL INFORMATION

■ **SIZE** The seas and oceans cover 362 million square km (140 million square miles), 71 per cent of Earth's total surface.

■ **AVERAGE DEPTH** 3,740m (12,270ft)

■ **GREATEST DEPTH** Marianas Trench, northwest Pacific, 10,915m (35,810ft)

■ **LARGEST OCEAN AND SEA** Pacific Ocean, 166 million sq km (64 million square miles), 46 per cent of the marine biome; average depth 3,950m (12,960ft). Coral Sea (off

northeast Australia), 4.7 million square km (1.8 million square miles) or South China (Nan) Sea, 3 million square km (1.1 million square miles), depending on definition.

■ **SMALLEST OCEAN** Arctic Ocean, 14 million square km (5.5 million square miles); depth 1,100m (3,610ft).

■ **SURFACE TEMPERATURE** From –3°C (37.4°F) in Beloya More (White Sea) off northwest Russia, to 37°C (98.6°F) in The Gulf.

193

COASTAL WATERS

COASTAL WATERS MAKE UP THE WORLD'S LONGEST, NARROWEST HABITAT: 844,230KM (524,266 MILES) OF COASTLINE THAT VARIES FROM SHEER ICE CLIFF AND BOULDER-STREWN SHORE TO PALM-FRINGED BEACH, HORIZON-WIDE MUDFLAT, AND CORAL REEF.

HIGH AND LOW COASTS
Shorelines are shaped by the land's geology, the water currents, and the prevailing winds that whip up waves. Very hard rocks, which resist much of the sea's erosion, stand as high cliffs. Softer rock is pounded into gently sloping beaches of sand, shingle, or pebble.

A RICH, DIVERSE HABITAT
Coastal waters extend from the intertidal zone (the shoreline itself), through the shallows, to the edge of the continental shelf – the stretch of seabed, 150–200m (500–660ft) deep, between the shore and the open ocean. At the edge of the continental shelf, the seabed suddenly falls away into deep ocean. The width of the shelf – and therefore the breadth of the coastal water – ranges from a few dozen metres to tens of kilometres.

Coastal waters are the most varied of the world's habitats. The relatively shallow water allows sunlight to penetrate as far as the seabed, and this – combined with nutrient run-off from land, currents that stir more nutrients from the seabed, and the very variable conditions – means that life in coastal waters is far richer and more diverse than it is in other marine habitats.

Many kinds of mammal visit the shore. Wolves, foxes, bears, hyenas, stoats, mink, raccoons, cats, and even monkeys and rats, dabble and paddle for meaty meals, grabbing fish and crabs from rock pools or scavenging washed-up carcasses. Various species of otter regularly swap their usual fresh water for the salt waters of the coast in order to hunt in the shallows. But the main group of mammals to divide their time between sea and land are seals, sea lions, and the walruses – the pinnipeds (or "flipper feet"), so called because their limbs have evolved into paddle-like flippers. The "leg" bones, which are short and stout, are inside the main body and form the flipper base. The digit (or toe) bones are elongated to form the main part of the flipper.

SEA LION
The male South American sea lion will posture on the beach at breeding time, barking and roaring at rivals.

BATTLE TO MATE   BEHAVIOUR

Most pinnipeds are colonial breeders. They gather at traditional sites, called rookeries, where the males (bulls) compete with each other for territories – small patches of sand or rock on the shore. The larger, stronger bulls win territories near the centre of the rookery, which is desirable since these central locations are the most attractive to breeding females.

TRAINING TIME
Rival young grey-seal bulls rear, shove, roar, and bite, as they learn to compete for territory.

Apart from the Baikal seal, found only in the fresh water of Russia's huge Lake Baikal, all pinnipeds are based in coastal waters. They use the land as somewhere to rest, groom their thick, waterproof fur, court, and breed; the sea is their hunting ground. However, some species, such as the massive elephant-seal, leave coastal waters after breeding and spend many months in the open ocean, resting and sleeping among the waves. The male southern elephant seal is the largest pinniped: a well-fed specimen weighs more than 5 tonnes (11,000lb) – as much as a real elephant.

OPPORTUNISTIC FEEDERS

Seals and sea lions are carnivores, and most hunt fish, holding their breath for several minutes as they descend to 50m (160ft) or more in search of a meal. Many species will take other prey, too - squid, clams dug from mud, mussels pulled from rocks, crabs, shrimps, and other crustaceans, as well as seabirds such as penguins and guillemots. A few of the larger species, such as Steller's sea lion of the North Pacific, hunt other mammals, such as otters and small pinnipeds.

There are three pinniped subgroups. The 14 species of eared seal (sea lions and fur seals) have small ear flaps; they can prop themselves up on the front flippers and use the rear ones to help them waddle over land. The 19 species of earless seals (true or hair seals) have no ear flaps, their flippers are useless out of water, so they move on land by wriggling or "humping" along on their bellies. The third group contains one species, the walrus, whose tusks make it very distinctive; its front flippers are like those of the sea lions, but the rear ones are more like those of the earless seals.

STELLER'S SEA LION   **COMMON SEAL**   **WALRUS**

"Life in **coastal** waters is far **richer** and more **diverse** than it is in other marine **habitats**."

BULLS ARE BIGGER
In most pinnipeds, such as these southern sea lions, the bulls (males) are much larger than the cows (females).

SURGING ONTO THE SHORE
In a wall of water, a 10-tonne (22,000-lb) killer whale "body-surfs" up the beach, grabs a hapless pup, and then twists around to drag its prey away into open ocean.

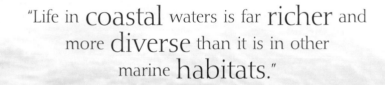

A TALE OF TWO MARINE MAMMALS
A newly weaned sea-lion pup must take to the water to hunt. But danger lurks just a few metres away, for killer whales know when it's sea lion breeding time.

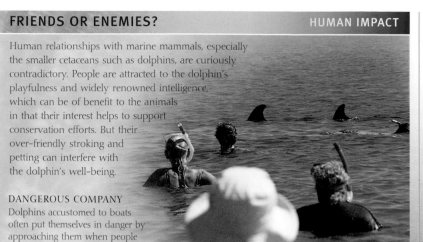

FRIENDS OR ENEMIES?

HUMAN IMPACT

Human relationships with marine mammals, especially the smaller cetaceans such as dolphins, are curiously contradictory. People are attracted to the dolphin's playfulness and widely renowned intelligence, which can be of benefit to the animals in that their interest helps to support conservation efforts. But their over-friendly stroking and petting can interfere with the dolphin's well-being.

DANGEROUS COMPANY
Dolphins accustomed to boats often put themselves in danger by approaching them when people are trying to land netted fish.

OTTERS OF THE SEA

Many otter species take to the sea occasionally, but two in particular are highly adapted to coastal life. Both are found in the Pacific. The marine otter dwells in the south Pacific along South America's western shorelines. It forages up to 500m (1,650ft) off shore, and sometimes enters rivers, but it is based on land. The sea otter of the north Pacific is not – it rarely touches dry land, even in rough weather. It is the most exclusively aquatic otter and, with a head–body length of 120cm (47in), a 20cm (8in) tail, and a bodyweight of 25kg (55lb), it is also the world's smallest marine mammal. There are three main sea otter populations: 2,000 Californian, 120,000 Alaskan, and (in the northwest Pacific) 20,000 Asian.

The sea otter is undeniably appealing to human eyes, with its frizzy pale fur, bleached by sun and salt, contrasting with beady eyes and black nose. These

ASLEEP ON THE OCEAN WAVE
Sea otters rarely come ashore. Using their rudder-like tails and flipper-like hind feet to keep afloat, they eat, rest, sleep, and groom themselves, at sea.

otters also display endearing behaviour: they manipulate items with their hands, habitually float on their backs as if sunbathing, and spend hours grooming and oiling their thick, water-repelling, heat-retaining fur. Although the female may give birth on rocks near the shore, she will immediately take her pup out to sea, carrying it on her belly to keep it out of the water.

Sea otters are social animals. They usually live in groups, called rafts, and frequent offshore "beds" or "forests" of kelp (a large, tough, broad-fronded brown seaweed). The floating fronds calm passing waves and support and shelter the otters as they rest (the otters will often anchor themselves to the fronds before sleeping). The kelp also provides a tangled mass of stems and bases among which the otters hunt for their meals.

SMASHING FOOD

Sea otters dive up to 30m (100ft), remaining under the water for as long as five minutes to bring up clams, abalones, crabs, spiny sea-urchins, and just about any other sea-floor animal. One of the otter's more remarkable habits is to bring up a small rock along with any prey that has a hard shell. The otter then lies on its back, puts the rock on its belly, holds the shell in its front paws, and smashes it open on the stone. It would appear that the otter is a fastidious diner, for after smashing open its food it swishes the meat in the water to remove mud and shell fragments.

STRANDED ON THE SHORE — BEHAVIOUR

Each year, hundreds of whales and dolphins are washed alive onto the shore. Common victims include pilot, killer, and sperm whales, and bottle-nosed dolphins. Sickness or old age may account for individual strandings. Mass strandings are more puzzling. Do they become lost in storms or unusual currents? Does sea pollution make them ill? Might underwater noise, such as that of ship propellers, interfere with their echolocation systems?

A RARE GLIMPSE
This Stejneger's, or Bering Sea, beaked whale is a deep-diving ocean-goer that is rarely seen. Much of our knowledge about such species is gleaned from stranded animals.

PROTECTION

In the 19th century the sea otter's appealing looks and habits did not prevent its mass slaughter, especially in the southern parts of its range – chiefly for its luxuriant fur but also because it raided shellfish beds cultivated by people. The otter is now a protected species and a much-loved inhabitant of the Californian coast. However, some other marine mammals – especially those that are more remote from our eyes because they live far out at sea – still face massive threats, such as continued hunting for meat and oil, pollution, and accidental drowning in fishing nets.

A FEAST AT SEA
A sea otter makes a feast of a crab, using its retractable claws to scrape the last remnants of meat from the shell.

OCEAN CONDITIONS

Mid-oceanic conditions are far more variable than they appear. Surface temperatures rise 10–15°C (18–27°F) under the midday sun, and there can be a similar difference between a cold current and a warm one (which may be just a kilometre or two apart). The strength and direction of the current fluctuates with the seasons and at different depths, while waves vary from imperceptible ripples to towering breakers.

OPEN SEA THE LARGEST HABITAT, AND ALSO

THE ONE LEAST UNDERSTOOD BY MAN, THE DEEP, MYSTERIOUS OCEAN SUPPORTS A DIVERSITY OF LIFE THAT RANGES FROM THE MICROSCOPIC ORGANISMS AT THE BOTTOM OF THE FOOD CHAIN TO THE GIGANTIC MAMMALS THAT ARE THE LARGEST ANIMALS ON THE PLANET.

A HUGE LARDER

In the world's greatest habitat, light penetrates no further than a depth of 250m (820ft), so life is concentrated near the surface in this sunlit (or photic) zone. Here, billions of tiny organisms make up the planktonic "soup" that nourishes the krill (small shrimp-like crustaceans), fish, and squid that in turn provide food for all open-ocean mammals. Krill is so plentiful that it sustains the world's largest animals.

A WHALE BLOWS ITS NOSE
The blowhole on top of a whale's head is its "nostrils". Water vapour in the warm expelled breath condenses into the steamy "blow" of this grey whale.

The whale and dolphin group, Cetacea, has two major divisions: the great or baleen whales (numbering 12 species) are filter-feeders; the toothed whales, which include dolphins and porpoises and number 71 species, are active hunters of larger prey. The baleen whales include the world's largest mammals: even the smallest species, the minke, exceeds 10m (33ft) and 12 tonnes (26,400lb). The largest is the blue whale at 30m (100ft) and 100 tonnes (220,000lb) or more.

PROFILE

GREAT WHALES

A great whale's body shape reflects its lifestyle and swimming speed. The wide and bulky right whale swims slowly and dives to only a few metres for a few minutes. Migrating super-sleek fin whales surge along for days at 25km/h (16mph).

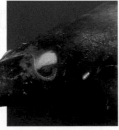

SOUTHERN RIGHT WHALE

FIN WHALE

TAKING A BREATH
A 70 tonne (154,000lb) fin whale – the largest species but for the blue – slides to the surface to breathe. Fin whales winter in subtropical waters (here, the Gulf of Mexico).

FLOPS, FLAPS, CLAPS, AND SLAPS
BEHAVIOUR

Great whales produce many and varied sounds – and not all are vocal. For example, many species display a behaviour known as breaching: they swim upwards with extraordinary speed and power in order to thrust their 50-plus tonnes (110,000lb) clear of the water and then flop back with a colossal splash. This breaching behaviour may help to get rid of barnacles and other external parasites, but it may also be used to send sound signals to other whales in the vicinity. Fluking (or lob-tailing), which involves lifting the tail (rather than the whole body) clear of the water and then slapping it onto the surface, may also be a form of communication.

BIGGEST SPLASH
(Above) A breaching humpback propels itself up before crashing back into the water. (Below) A southern right whale prepares to fluke.

BALEEN WHALES

Baleen whales are named after the substance baleen, or "whalebone", which enables them to filter-feed. Baleen is not bone, but a strip or plate of a flexible, springy substance something like stiff plastic; it has stringy fringes that hang from the whale's upper jaw, looking like a massive, curved, brush-edged curtain. Each great whale species has its own number, size, and design of baleen plate. The largest belong to the bowhead whale, which has more than 600 plates, each of which is over 4m (13ft) long and 15–25cm (6–10in) wide. The blue whale has up to 400 plates, each of which is less than 1m (3ft) long.

SIEVED FROM THE SEA

To feed, a typical baleen whale gapes wide and takes in a gigantic mouthful of water, ballooning its pleated chin. Then the mouth almost closes and water is forced out through the plates and between the lips. Small food items such as krill or fish are trapped in the baleen sieve, and then licked off by the whale's car-sized tongue. During a day's filter-feeding, a blue whale consumes 5 tonnes (11,000lb) of krill.

"The baleen whales include the world's largest mammals; the largest of all is the blue whale at 30m (100ft) and 100 tonnes (220,000lb) or more."

WHALE-WATCHING
HUMAN IMPACT

One of the fastest-growing wildlife activities is taking an organized boat trip to observe whales, especially as they feed during summer in the waters of the far north and south. Visitor payments help to support the local economy; this in turn encourages whale conservation so that the spectacle can continue to attract funds in the future.

LUNCH-TIME IN ALASKA
Whale-watching vessels keep their distance, both for their own safety and to avoid disrupting the mammals' activities.

A HUMPBACK SURFACES AT SPEED
The whale's nose points skywards, showing the gaping upper jaw, roof of the mouth, and curved rows of baleen plates. The vast chin and throat below the lower jaw is hugely distended by a watery mouthful of krill.

SHORT-FINNED PILOT WHALES
At 7m (23ft) long and over 1 tonne (2,200lb), pilots are among the largest of the dolphins. Like many toothed species, they stay close and share food.

KEEPING IN TOUCH
Common dolphins dive down to 300m (1,000ft) in pursuit of shoaling fish and squid, filling the sea with the clicks and squeals that serve to echolocate prey and to inform each other of their positions.

DOLPHINS

All members of the dolphin family are streamlined, rapid swimmers, leaping clear of the surface at speed for various reasons: to check on their whereabouts and other pod members, to escape predators such as sharks, and to locate or herd prey shoals.

This action is known as "porpoising", although porpoises rarely do it. Dolphin sizes range from Commerson's dolphin, only 1.5m (5ft) and 80kg (175lb), to the similarly coloured killer whale, six times longer and 100 times heavier.

DALL'S PORPOISE

AMAZON RIVER DOLPHIN

COMMERSON'S DOLPHIN

TOOTHED WHALES OF THE WORLD

Toothed whales, or odontocetes, form a group that has six times more species than the baleen group has. The toothed whale group includes the only freshwater cetaceans (five species of river dolphin), porpoises (six species), two white whales (beluga and narwhal), sperm whales (three species), the deep-diving and rarely seen beaked whales (19 species), and the dolphin family (33 species). All are active, fast-swimming predators. Harbour porpoises nose the inshore seabed for fish and shellfish, while dolphins speed through surface waters after rapidly moving shoals of fish and squid; beaked whales make long, deep dives in the open ocean to find crabs, starfish, and urchins.

Almost all toothed whales are social mammals. They live in smallish (generally less than 20), relatively stable, single-species groups called pods. Members communicate using various sounds and body postures; they form long-term relationships with each other, and they often hunt co-operatively. Pods may merge

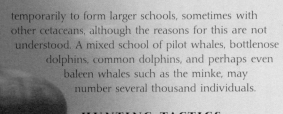

temporarily to form larger schools, sometimes with other cetaceans, although the reasons for this are not understood. A mixed school of pilot whales, bottlenose dolphins, common dolphins, and perhaps even baleen whales such as the minke, may number several thousand individuals.

HUNTING TACTICS

Many toothed whales travel so fast and dive so deep that their hunting methods in the wild are difficult to document. The most intensively studied species are those in captivity: the bottlenose dolphin, and the killer whale or orca (the largest dolphin species). Both exhibit their intelligence by learning and even inventing their own tricks. In the wild, the killer whale pod is made up of an extended family led by a senior female, or matriarch. Pod members communicate with whistles, clicks, squeals, and grunts, and they co-ordinate hunting tactics in many ways: they may surround a spread-out shoal of fish and swim inwards to compress it into a tight ball, to make feeding easier, or they may herd the shoal into a shallow bay with no escape. In some regions killers ride the surf onto the shore, intentionally beaching themselves in order to grab unwary sea lions or seals (see page 194-95), or they flip over small ice floes to tip these prey into the water.

PARASITIC PASSENGERS
Many cetaceans – especially the slower-swimming baleen species such as this grey whale – become encrusted with parasites such as barnacles, tube-worms, and types of crab-like crustaceans called "fish-lice".

GIANT HUNTERS
Sperm whales – the largest carnivorous predators on the planet – often rest close to the water's surface, stroking other pod members with their flukes and flippers.

FROM LAND TO SEA – HOW WHALES HAVE EVOLVED

EVOLUTION

After the death of the dinosaurs and the large sea reptiles 65 million years ago, mammals evolved very rapidly, some making their way into the sea. One of the first known cetaceans, from 50 million years ago, was Pakicetus, whose fossils were found in Pakistan. It probably evolved from an early ungulate (hoofed mammal), which first switched to a carnivorous way of life and then began to enter the water. By 35 million years ago, Basilosaurus, whose fossils come from eastern North America, showed complete aquatic adaptation, with front limbs modified as flippers, rear limbs lost, and wide flukes at the end of the tail.

PAKICETUS
Pakicetus was a "halfway whale", with paddle-like limbs for moving on land as well as in water. It reached 1.8m (6ft) in length.

SPERM WHALE
The cetacean most adapted to deep-sea life is the sperm whale: it can dive below 2,000m (6,550ft) and stay under for more than two hours.

BASILOSAURUS
This fully ocean-going whale, 25m (82ft) in length, was more elongated and eel-like than today's toothed whales.

POLAR HABITATS
Stony-shored islands in subpolar waters provide haul-out sites for seals and sea lions, but in winter they are iced over and too cold to support other life. Closer to the poles, huge icebergs continually crack (or "calve"), creating massive glaciers and ice shelves.

POLAR
THE ARCTIC AND ANTARCTIC POLES ARE VERY DIFFERENT: THE FORMER IS A SHALLOW OCEAN ALMOST SURROUNDED BY LAND; THE LATTER IS A GIANT CONTINENT WITH DEEP SEA ALL AROUND IT. BUT BOTH HAVE A BRIEF SUMMER WHEN LIFE FLOURISHES IN THE WATER AND SUPPORTS AN ABUNDANCE OF MARINE MAMMALS.

EXTREME CONDITIONS

Nowhere on Earth are the seasons more extreme than they are in the north and south poles. When the six months of perpetual darkness in winter give way to spring's rapidly extending daylight, the tiny plants of the plankton bloom in the cool but nutrient-rich waters. Small creatures feast on the harvest and are food for fish, krill, squid, and other small predators, which are in turn consumed by some of the world's largest meat-eaters.

Many kinds of great whale (pp.198–203) migrate thousands of kilometres to gorge themselves on the 4-month polar summer harvest before they cruise back to warmer but nutrient-poor subtropical waters for the winter fast. Smaller whales, and numerous seals and sea lions, undergo shorter journeys of several hundred kilometres. They follow the ice's shrinking edge polewards in spring, and return in autumn as the glaciers, sheets, and icebergs spread once again.

About half of the 34 species of pinniped (seals and sea lions) are found in polar or subpolar waters for at least part of the year. They are superbly adapted to withstand the cold, with dense fur coats and a thick layer of insulating fatty blubber under the skin. The blubber also helps to round out body contours, which makes the animal better streamlined for swimming; and when prey is scarce the blubber acts as an emergency food store. Even so, maintaining a warm body temperature in near-freezing temperatures, while still having sufficient stamina to dive after speedy prey, demands a plentiful high-energy diet. A seal's diet is 100 per cent flesh and, in relation to body weight, the food intake is five times greater than that of a human.

SURVIVING ON ICE

Seals and sea lions must leave the water to reproduce. In summer the small islands, ice shelves, and icebergs of Antarctic waters are alive with male pinnipeds competing for females, and females suckling their pups. The most southerly-breeding species is the crabeater seal, which rears its pups on the pack ice fringing Antarctica itself. With an estimated total population of 15 million, the crabeater is probably the world's most abundant larger wild animal. It is also,

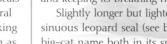

A BREATH OF FRESH AIR
Should one of its breathing holes freeze over, the Weddell seal can restore air access by ramming and biting through fresh ice 20cm (8in) thick.

perhaps, one of the most misnamed, since there are no crabs in the crabeater's habitat. Instead this seal uses its unusual lobe-shaped teeth to sieve the water for krill. Like most seals, the crabeater dives to about 50m (160ft) and remains under water for several minutes. A much more proficient diver is the Weddell seal (see left), which can reach depths of 500m (1,650ft) and stay submerged for more than an hour in search of fish, squid, and other prey. As sea ice spreads in the winter darkness, the Weddell dives through holes in the ice to hunt, but must then remember the site of the hole so that it can come up to breathe. It divides its time between hunting and keeping its breathing holes clear of fresh ice.

Slightly longer but lighter, at 350kg (770lb), is the sinuous leopard seal (see below left). It lives up to its big-cat name both in its spotted appearance and in its hunting prowess: its sizeable prey includes penguins, albatrosses, as well as other seals.

TIME FOR A REST
Named after his trunk-like nose and vast bulk, the male elephant-seal takes a break from roaring and biting rivals.

FAST AND DEADLY
Two-fifths of the leopard seal's diet is warmblooded - other seals, birds, and chunks bitten out of ill or injured great whales. The rest is made up of krill, fish, squid, and shellfish.

FENCING CONTESTS   EVOLUTION

The male narwhal's upper left incisor tooth has evolved into a spike-like tusk. Up to 3m (10ft) long, it grows through the lip, spiralling forward and lengthening with age. It is not vital for feeding, for prey is sucked in by the powerful lips and tongue – and nearly all females are tuskless. Rather, it is used like a sword at breeding time to intimidate rival males during "fencing" contests.

TRULY ARCTIC
The narwhal is the most permanently northern of all mammals, feeding among the floes that edge the Arctic ice sheet.

SAVING ENERGY

Crabeater seals haul out onto an ice shelf to rest.
In the water they hunt for the small shoals of fish
and krill that hide under the ice.

WHISKERS AND TUSKS

Third-largest pinniped (after the southern and northern
elephant-seals), the 2-tonne (4,400-lb) walrus inhabits
the shores and shallows around the Arctic Ocean and
the northern Pacific and Atlantic. The front flippers
resemble a sea lion's and can therefore be used to
prop up the front half of the body while on land.
The rear flippers are more like a seal's, so while they
are very effective in water they are useless on land.
As with other pinnipeds, the walrus is well

adapted to the icy conditions of its habitat. In addition to insulating blubber, this massive beast has heavily creased skin that is up to 4cm (1½in) thick and sparsely furred with hairs about 1cm (⅖in) long; on the blunt muzzle the hairs sprout as long, thick whiskers. The skin is greyish brown, but when the animal rests in the sun the blood vessels dilate to enhance heat absorption, and the skin then changes to deep pinkish red. In colder conditions, and in the water, the blood vessels contract again to reduce heat loss.

The walrus is classified in its own pinniped subgroup, Odobenidea, which means "tooth-walker". The name was bestowed following 18th-century Arctic explorers' tales that walruses moved on land by extending the neck, jabbing the long tusks into the ice, and using them as a pick to haul the body forwards. Both sexes have tusks, which are elongated upper canine teeth that grow with age. The tusks do not show beyond the lips until the animal is approaching two years of age, and the female's tusks tend to be shorter, slimmer, and more curved than they are in males of the same age. Older bulls have the longest tusks – up to 1m (3ft).

SYMBOLS OF STATUS

Walruses sometimes use the "ice-pick" technique to pull themselves out of the water at the ice's edge. Other tusk uses include self-defence against polar bears and killer whales, and breaking holes in the ice through which they dive to depths of up to 100m (330ft) in search of food. The tusks may

WATERY TERRITORY   BEHAVIOUR

Walrus bulls battle to establish small aquatic territories just off the shore. These are used as bases from which to call to passing females. Disputes between equally matched bulls may escalate from noisy posturing and symbolic jabbing movements, to attempts at actual stabbing of the opponent, although fatalities are rare. Cows occasionally have similar but less intense disputes.

DAGGERS DRAWN
The shoulders of older males develop very thick skin, which is scarred by years of battle.

also be used for levering food from the seabed. But their foremost function is to show age and status. Walruses are gregarious and rest in large herds that are sometimes hundreds strong. The individuals with the largest tusks assume dominance, and therefore have the best resting places as well as access to mates during the three months of the breeding season.

Bulls do not become sexually mature until the age of about ten, but competition is fierce and they do not usually attain sufficient body bulk and tusk size to win a mate until they are about 15. Cows, which are around half the weight of bulls, usually have their first pup when they are six or seven. Pups are weaned at about 18 months. Walrus courtship is a noisy affair. Whenever the females move into the sea to feed, the males entice them with loud, long underwater calls that include whistles, bell-type clangs, knocks, and handclap-like sounds.

CROWDED BEACH
The typical walrus group contains several females with young. They form a loose herd when feeding at sea, but merge with other groups to sunbathe on shore. Bulls wander among them.

A REST BEFORE FEEDING?
A young walrus rests chin on foreflipper. When he is hungry he will search the seabed for animals such as worms, shellfish, and bottom-dwelling fish, locating them with his sensitive snout and long whiskers.

GLOSSARY

A

ANTICOAGULANT Any substance that, when added to blood, prevents it from clotting.

ANTLER A bony growth on the head of deer. Unlike horns, antlers often branch, and in most cases they are grown and shed every year in a cycle linked with the breeding season.

AQUATIC Living wholly or partly in water.

ARABLE Relating to the use of land to grow crops.

ARBOREAL Living wholly or partly in trees.

B

BINOCULAR VISION Vision in which the two eyes face forwards, giving overlapping fields of view. This allows an animal to judge depth.

BIOME A characteristic group of living things, together with the setting in which they are found.

BOVID Any hoofed mammal in the order Artiodactyla. This includes antelopes and buffaloes, cattle, sheep, and goats.

BROWSING Feeding on the leaves of trees and shrubs, rather than on grasses. See also Grazing.

C

CAMOUFLAGE Colours or markings that enable an animal to merge with its background. It is used both for protection against predators, and for concealment when approaching prey. See also Cryptic coloration.

CANID A member of the family Canidae, a group of carnivores that includes wolves and dogs.

CANINE TOOTH A tooth with a single sharp point designed for piercing and gripping. Canine teeth are located either side of the incisors and are highly developed in carnivores.

CARNASSIAL TOOTH Very strong, deep-rooted, blade-like premolar in carnivorous mammals designed for slicing flesh and breaking bone.

CARNIVORE Any animal that eats meat. Also used in a more restricted sense to mean animals of the order Carnivora.

CHAPARRAL Vegetation composed of broadleaved evergreen shrubs, bushes, and small trees usually less than 2.5m (8ft) tall; together they often form dense thickets. Chaparral is found in regions characterized by hot dry summers and mild, wet winters.

COLDBLOODED See Ectothermic.

CONIFER Any member of the order Confederales: woody plants that bear their seeds and pollen on separate, cone-shaped structures.

CRUSTACEAN A member of the class Crustacea, a group of invertebrate animals consisting of some 39,000 species distributed worldwide. Crabs, lobsters, and shrimps are examples of crustaceans.

CRYPTIC COLORATION Coloration and markings that make an animal difficult to see against its background.

D

DASYURID Any member of the family Dasyuridae, a group of marsupial mammals that includes the Australian native cats, marsupial mice and rats, and their allies.

DECIDUOUS Of trees (primarily broadleaved) that shed all their leaves during one season.

DIURNAL Active during the day.

DOMESTICATED Of an animal that lives fully or partly under human control. Many have been bred to produce artificial varieties, which are not found in the wild. See also Feral.

DORSAL On or near an animal's back.

DREY The nest of a squirrel, which is built in a tree.

E

ECHOLOCATION A method of sensing nearby objects by using pulses of high frequency sound. Echoes bounce back from obstacles and other animals, allowing the sender to build up a "picture" of its surroundings. Used by several groups of animals, including bats.

ECOSYSTEM A unit consisting of a habitat and the collection of species that live within it.

ECTOTHERMIC Having a body temperature that is dictated chiefly by external conditions. Also known as coldblooded.

ENDOTHERMIC Able to maintain a constant, warm body temperature, regardless of external conditions. Also known as warmblooded.

F

FAMILY A level used in classification. In the sequence of classification levels, a family forms part of an order and is subdivided into one or more genera (see Genus).

FERAL Relating to an animal that comes from domesticated stock but has subsequently taken up life in the wild. Animals that most commonly become feral include cats, city pigeons, and horses.

FILTER–FEEDER An animal that feeds by sieving small food items from water. Examples include baleen whales.

FJORD A long narrow arm of the sea, commonly extending far inland, that results in marine inundation of a glacial valley.

FLIPPER In aquatic animals, a paddle-shaped limb.

FLUKE A rubbery tail flipper in whales and their relatives. Unlike the tail fins of fish, flukes are horizontal and beat up and down.

FOOD CHAIN A food pathway that links two or more different species in that each forms food for the next species higher up in the chain. In land-based food chains, the first link is usually a plant. In aquatic food chains, it is usually an alga or other form of single-celled life.

FORM A depression in the ground used by hares for concealing themselves or their young.

G

GENUS (pl. genera.) A level used in classification. A genus forms part of a family, and is subdivided into one or more species.

GESTATION PERIOD The period between fertilization and birth, when the developing young is nourished by the mother via the placenta.

GONDWANA Hypothetical former supercontinent in the Southern Hemisphere, which included modern South America, Africa, Australia, India, and much of Antarctica.

GRAZING Feeding on grass. See also Browsing.

GUARD HAIR A long hair in a mammal's coat, which projects beyond the underfur and helps to keep it dry.

H

HERBIVORE An animal that eats only plants or plant-like plankton.

HIBERNATION A period of dormancy in winter. During hibernation, an animal's body processes slow up, conserving energy and reducing the requirement for food.

HORN A hard, pointed growth on the head. Unlike antlers, horn is hollow and made up of keratin.

I

INCISOR TOOTH A tooth at the front of the jaw that is shaped for slicing or gnawing.

INSECTIVORE An animal that feeds on insects. In a more

restricted sense, it means animals of the mammalian order Insectivora.

INTRODUCED SPECIES A species accidentally or deliberately brought into an ecosystem in which it does not naturally occur.

INVERTEBRATE An animal without a backbone.

L

LEK A communal display area used by male animals, such as deer, during courtship.

M

MARSUPIAL Any member of Marsupialia, a mammalian order characterized by premature birth and continued development of the young while attached to the nipples on the belly of the mother.

MEANDER Extreme U-bend in a stream, usually occurring in a series, that is caused by the flow characteristics of the water.

MIGRATION Annual or seasonal journeying from one region to another, usually in order to take advantage of good breeding conditions or feeding opportunities.

MOLAR TOOTH A deep-rooted tooth at the rear of the jaw. Used for chewing, it usually has a flat or ridged surface.

MONOTREME Any member of the egg-laying mammalian order Monotremata, which includes the platypus and the echidnas of Australia.

MORPHOLOGY Shape and structure; the form of an animal or its bodyparts

N

NATIVE Having been born in an area, or having naturally occurred there for a very long time. Compare Introduced.

NOCTURNAL Active at night.

O

OMNIVORE An animal that eats both plant and animal food.

OPPOSABLE Able to be pressed together from opposite directions. An opposable thumb - a feature that is found in many primates - can be pressed against the other fingers.

ORDER A level used in classification. An order forms part of a class, and is subdivided into one or more families.

P

PAMPAS Vast, grass-covered plain in South America, extending west from the Atlantic coast to the Andean foothills, primarily in Argentina.

PATAGIUM In bats, the flap of double-sided skin that forms the "wing". This term is also used for the parachute-like skin flaps of colugos and other gliding mammals.

PERISSODACTYLS A hoofed mammal with an odd number of toes; includes horses, tapirs, and rhinoceroses. Artiodactyls are even-toed, and include camels, deer, and cattle.

PLACENTA An organ developed by an embryo animal that allows it to absorb nutrients and oxygen from its mother's bloodstream before it is born. Young that develop in this way are called placental mammals.

PLACENTAL MAMMAL See Placenta.

POLYGAMOUS Of a reproductive system in which individuals mate with more than one partner during the course of a single breeding season.

POLYGYNOUS Of a polygamous male.

PREDATOR An animal that catches, kills, and eats other animals (which are called its prey). Some predators catch their prey by lying in wait, but most adopt an active pursuit and attack strategy.

PREHENSILE Of a limb or body part that is able to curl around objects and grip them.

PREY Any animal that is eaten by a predator.

PROBOSCIS An animal's nose, or a set of mouthparts with a nose-like shape.

R

RANGE 1) To roam or travel long distances in search of food, shelter, or mates. 2) An area used regularly by an animal for feeding, shelter, and breeding.

REFUGIA Pre-glacial habitats or areas supporting animal life that has adapted to their specific conditions.

RUMINANT A hoofed mammal that has a specialized digestive system, with several stomach chambers. This usually requires the food to be regurgitated and rechewed, a process known as "chewing the cud".

RUTTING SEASON In deer, a period during the breeding season when males clash with each other for the opportunity to mate.

S

SAVANNA Type of grassland found in the tropics; it is usually sparsely dotted with deciduous trees.

SPECIES A group of similar organisms that are capable of interbreeding in the wild, and of producing fertile offspring that resemble themselves. Species are the fundamental units used in biological classification. Some species have distinct populations that vary from each other. Where the differences are significant, and the populations biologically isolated, these forms are classified as subspecies.

STEPPE Area of treeless grassland in which short, drought-resistant grasses predominate. (The term is most often used to describe the belt of grassland that extends 5,000 miles (8,000km) from Hungary in the west, through Ukraine and central Asia, to Manchuria in the east.)

SUBCLASS A level used in classification, between a class and an order.

SUBFAMILY A level used in classification, between a family and a genus.

SUBORDER A level used in classification, between an order and a family.

SUBSPECIES See Species.

SUBTROPICAL Nearly tropical, bordering the tropical zone. See also Tropical, Temperate.

SYNAPSID Any member of the extinct suborder Synapsida, mammal-like reptiles of the Triassic period.

T

TEMPERATE Moderate climate characterized by long, warm summers and short, cold winters.

TERRESTRIAL Living wholly or mainly on the ground.

TERRITORY An area defended by an animal, or group of animals, against members of the same species. Territories often include useful resources that help the male attract a mate.

TROPICAL Climate characterized by high temperature, humidity, and rainfall. It is found in a belt that extends both sides of the equator, between the Tropics of Cancer and Capricorn.

TUSK A modified tooth that often projects outside the mouth. Tusks have a variety of uses, including defence and digging up food. In some species, only the males have them - in this case their use is often for sexual display.

U

UNDERFUR The dense fur that is closest to a mammal's skin. Underfur is usually soft and a good insulator.

UNGULATE A hoofed mammal.

V

VERTEBRATE An animal with a backbone.

W

WARMBLOODED See Endothermic.

WEANING The period when the mother gradually ceases to provide milk for her young.

INDEX

ACKNOWLEDGMENTS

Packager's Acknowledgments
Many thanks to Jo Weeks for proofreading, Hilary Bird for indexing, Jon Hughes for illustrations, and Rob Stokes for map artworks. Thanks to Sue Gordon, Aaron Brown, Tom Butler, Emily Hawkins, and Abbey Cookson-Moore for editorial assistance. Special thanks to Mic Cady for his boundless energy and enthusiasm throughout the project, and for his invaluable organizational skills during the crucial early stages. Thanks also to the DK, OSF, and Smithsonian teams, and particularly to the unflappable Angeles Gavira and Ina Stradins. Finally, special thanks to Dr George C. McGavin.

Picture Credits
The publisher would like to thank the following for their kind permission to reproduce their photographs;

(Abbreviations key; t = top, b = below, r = right, l = left, c = centre, f = far, a = above)

(Agency abbreviations key; Oxford Scientific Films: OSF)

1: OSF/Mark Jones (c); **2–3:** Getty Images/ Beverly Joubert; **4:** Corbis/Randy Wells (crb); **4:** OSF/Rafi Ben-Shahar (cra); **5:** OSF/Clive Bromhall (clb), Frithjof Skibbe(tr), L.L Rhodes/ AA (tl); **5:** Royalty Free Images/ Corbis (cla), Getty/Digital Vision (crb); **8–9:** OSF/Richard Packwood; **10:** Corbis/Charles Philip (cla); **10–11:** FLPA -Images of nature/Minden Pictures/ Flip Nicklin **10:** OSF/Cynthia Moss (cfl), Mark Stouffer/AA car; Thomas Haider (b); **11:** OSF/ Mike Hill (c), Raymond Mendez (tcl); **12:** alamy.com/ Steve Bloom (bl); **12:** OSF/Stan Osolinski (tr), (tcr); **13:** OSF/David Cayless (cra), Gerard Soury (br), Mantis Wildlife Films (bl), OKAPIA (bc), OSF (crb); **14:** OSF/Claude Steelman/SAL (cr), Des & Jen Bartlett/SAL (bc); **14–15:** OSF/Norbert Rosing; **15:** OSF/John Downer (bcl), Mark Hamblin (bcr), Martyn Colbeck (bc), Stan Osolinski (br), Werner Pfunder (c); **16:** Corbis/Tom Brakefield (cla); **16:** N.H.P.A./John Shaw (br); **16:** OSF/Daniel Cox (clb), Eyal Bartov (tl), Nick Gordon (br), Richard Herrmann (clb), Stan Osolinski (bl); **17:** Corbis/John Conrad (b); **17:** OSF/Brian Kenney (ca), Daniel Cox (cra), Michael Leach (tl), Owen Newman (crb); **19:** Getty Images/John Giustina. **20:** Corbis/Joe McDonald (cfl); **20:** DK Picture Library/Hunterian Museum (tr); **20–21:** Getty Images/John Downer; **22:** OSF/Ben Osborne (c), Nigel Smith/AA (ca), Paul Franklin (b); **23:** OSF/C. C. Lockwood/AA (c), Charles Palek (ca), Colin Monteath (tr), Konrad Wothe (br); **24:** Ardea London Ltd/Pat Morris (bcl); **24:** Fotomedia/E. Hanumantha Rao (cra); **25:** OSF/K & L Laidler (cal); **26–27:** OSF/Rafi Ben-Shahar; **28:** Corbis/Gallo Images (br); **28:** OSF/McKinnon Films Ltd. (cr); **29:** OSF/Gallo Images (cra), Martyn Colbeck (cfr), Rob Nunnington (bc); **30:** OSF/Peter Lack (tl), Rafi Ben-Shahar (ca), Stan Osolinksi (cla), Stan Osolinski (cfl); **30–31:** Rob Nunnington (c); **31:** Peter Cross (tr); **32:** OSF/ Norbert Rosing (cla), Stan Osolinski (tl); **32–33:** Tim Jackson; **33:** Ardea London Ltd/Pat Moris (br); **33:** Brock Fenton (tr); **33:** N.H.P.A./Anthony Bannister (crb); **33:** OSF/Daniel Cox (tcr), Steve Turner (tc); **34–35:** OSF/David McDonald; **36:** Chris Mattison Nature Photographics/Martin Withers (tc); **36:** OSF/Rafi Ben-Shahar (br); **37:** Corbis/Jim Zuckerman (br); **37:** OSF/ Johnny Johnson (t), Stan Osolinki (crb); **38:** Steve Bloom/stevebloom.com; **39:** Corbis/Chris Hellier (bc); **39:** Peter Cross (tr); **39:** Nigel Dennis (tc), (tcl); **39:** Chris Mattison Nature Photographics/Martin Withers (tcr); **39:** OSF (br), Alan Root (cfr), Nick Gordom (ca); **40:** OSF/Berndt Fischer (ca); Michael Fogden cfl; Richard Packwood cla; Stan Osolinski tl. **40–41:** McKinnon Films Ltd (b); **41:** Ardea London Ltd/Chris Harvey (cra); **41:** Nigel Dennis (cfr); **41:** Brock Fenton (tr); **41:** Mike Jordan (crb); **41:** OSF/Michael Fogdon (br), Tim Jackson (tc); **42:** Corbis/Tony Wilson-Bligh; Papilio (b); **42:** Peter Cross (tl); **42:** OSF/Alain Dragesco-Joffe (cl), Mark Deeble & Victoria Stone (cra); **43:** OSF/Richard Packwood **44:** OSF/Barbara Wright/AA (bl), David Cayless (cla), Doug Allan (cfl), Richard Packwood (tl), Stan Osolinski (c); **45:** Corbis/Gallo Images (br); **45:** OSF/Adam Jones (tr), Mark Deeble & Victoria Stone (tc); **46–47:** OSF/Anthony Bannister (b); **47:** Ardea London Ltd/Pat Moris (br); **47:** Nigel Dennis (cfr); **47:** OSF/Jen & Des Bartlett (tr), Owen Newman (tc); **48:** OSF (car), Partridge Films Ltd. (tl), Tony Bomford (cla); **48–49:** Gallo Images (c); **49:** OSF/Adrian Bailey (cra), Bill Paton/SAL (cla), Konrad Wothe (crb); **50–51:** OSF/Mark Deeble & Victoria Stone; **52–53:** OSF/Anthony Bannister; **52:** OSF/Gallo Images (tl), John Downer (bc), Mary Plage (cfl); **54:** OSF/ John Dower (cla), Nick Gordan (cr), Richard Packwood (cfl), Steve Turner (tl); **54–55:** Martyn Colbeck (b); **55:** Ardea London Ltd/Alan Weaving (tcr); **55:** Bruce Coleman Ltd/Rod Williams (tc); **55:** DK Picture Library/Jerry Young (tl); **55:** Brock Fenton (tr); **55:** Nature Picture Library Ltd/Jeff Foott (cla); **55:** OSF/ Stan Osolinksi (c); **56–57:** Corbis/Gallo Images; **56:** S V Den Nieuwendjik (tr); **56:** OSF/Konrad Wothe (bc), Mike Birkhead (c); **56:** Art Wolfe (cla); **58:** OSF/Ken Cole/AA (bl); **58:** Kevin Schafer; **59:** Martin Harvey Photography (cr); **59:** FLPA - Images of nature: Terry Whittaker (br); **59:** OSF/Gallo Images (cra), Richard Packwood (ca), Robert Harvey/SAL (bl); **60:** Bruce Coleman Ltd/Rod Williams (c); **60:** OSF/David Haring (cla), Mike Powles (tc), (bc); **61:** Nigel Dennis (clb); **61:** Nature Picture Library Ltd/Alain Compost (cfl); **61:** OSF/Joe Blossom/SAL (bc), Mark Pidgeon. **61:** Dave Watts (cla); **62–63:** Corbis/Randy Wells. **64:** OSF/Daniel Cox (br), Tom Ulrich (cr); **65:** Alamy (br); **65:** Corbis/Farrell Grehan (cra); **65:** OSF/Judd Cooney (cb); **66:** OSF/Daniel Cox (bc), John Gerlach (cfl), Mary Plage (ca), Professor Jack Dermid (cla), Stan Osolinksi (tl); **67:** Alamy (b); **67:** OSF/Alan & Sandy Carey (ca), Wendy Shattil and Bob Rozinski (cra); **68:** OSF/Judd Cooney (tcl); **68–69:** Getty Images/Paul and Lindamarie Ambrose (b); **69:** OSF/Stan Osolinski (tr); **70:** N.H.P.A/ Stephen Krassemann (cfl); **70:** OSF/Alan & Sandy Carey (c), Claude Steelman/SAL (clb), Raymond Mendez/AA (bl), Wendy Shattil and Bob Rozinski (tl); **71:** Corbis/Westmorland (br); **71:** OSF/James H. Robinson (cla), Scott Camazine (cfr); Stan Osolinski (tr); **72:** OSF/ Alastair Shay (tl), Daniel Cox (b), Michael Fogden (cla); **73:** Mary Clay (cfr); **73:** Peter Cross (cra); **73:** OSF/Judd Cooney (br), Mary Plage (tc); **74:** OSF/Claude Steelman/SAL (cl), Michael H. Franci/OKAPIA (tl), Ronald Toms (c), Tom Ulrich (tc); **74:** Corbis (bc); **75:** Corbis/Sanford/Agliolo (c); **76:** OSF/Judd Cooney; **77:** OSF/Henry Ausloos (c), Jack Wilburn (tr), Richard Packwood (br), Stan Osolinksi (cra); **78:** OSF/Daniel Cox tr; Erwin & Peggy Bauer (br), Mike & Elvan Habicht (bl), Wendy Shattil and Bob Rozinski (tl); **79:** OSF/Alan Nelson (br), Wendy Shattil and Bob Rozinksi; **80–81:** Corbis/Farrell Grehan; **80:** OSF/Daniel Cox (cr), (clb), Michael Fogden (cfl), Stan Osolinski (cla), Zig Leszczynski (tl); **81:** Corbis/Chase Swift (br); **81:** OSF/Breck P. Kent (tc), Brian Kenney (crb), Norbert Rising (cra); **82:** OSF/C. C. Lockwood (cra), Daniel Cox (bl), Joe McDonald (tl); **83:** OSF/Alan & Sandy Carey (tr), Liz Bomford (tc), Ralph A. Reinhold (b); **84–85:** OSF/Stouffer Enterprises Inc; **86:** Robert E Barber (tr); **86:** William Bernard Photography/William Bernard (tcr); **86:** FLPA - Images of nature/S & D Maslowski (tcl); **86:** OSF/Daniel Cox (c), Mike & Elvan Habicht (br); **87:** OSF/Richard Alan Wood; **88:** OSF/ Frank Huber (bc), Tom Ulrich; **89:** OSF/E. R Degginer (b), OKAPIA (cra), Ronald Toms (c); **90:** OSF/Lon E. Lauber (c), Richard Kolar (cla), Stan Osolinski (tl); **90–91:** Tom Ulrich (b); **91:** OSF/Alan & Sandy Carey (tl), Leonard Lee Rue (cla), Mark Hamblin (tr), Richard Packwood (cbr), Rick Price (crb), Ted Levin (cfr); **91:** Andrey Zvoznikov (br); **92:** OSF/Bennett Productions/ SAL (bl), Daniel and Julie Cox (cal); **92–93:** Norbert Rosing; **93:** OSF/Dan Guravich (tc), Daniel Cox (c), Sue Flood (bc), (br); **95–95:** OSF/ L.L Rhodes; **96:** OSF/Chris Catton (crb), Konrad Wothe (cfr); **97:** OSF/Michael Fogden (tc), Rick Price/SAL (crb), Wendy Shattil & Bob Rozinski (cra); **98:** OSF/Berndt Fischer (tl), Colin Monteath (cla); **98–99:** Wendy Shattil & Bob Rozinski; **99:** OSF/M Wendler/OKAPIA (tr), Michael Dick/AA (cla); **100–101:** Corbis/Tom Brakefield (l); **101:** Corbis/Martin Harvey; Gallo Images (tr); **101:** OSF (br), Alan Root/SAL (cfl); **102:** OSF/Chris Sharp (tl), Des & Jen Bartlett/ SAL (bl), Judd Cooney (c), Richard Packwood

ACKNOWLEDGMENTS

OSF/Daniel Cox (cr), (clb), Michael Fogden (cfl), Stan Osolinski (cla), Zig Leszczynski (tl); **81:** Corbis/Chase Swift (br); **81:** OSF/Breck P. Kent (tc), Brian Kenney (crb), Norbert Rising (cra); **82:** OSF/C. C. Lockwood (cra), Daniel Cox (bl), Joe McDonald (tl); **83:** OSF/Alan & Sandy Carey (tr), Liz Bomford (tc), Ralph A. Reinhold (b); **84–85:** OSF/Stouffer Enterprises Inc; **86:** Robert E Barber (tr); **86:** William Bernard Photography/William Bernard (tcr); **86:** FLPA - Images of nature/S & D Maslowski (tcl); **86:** OSF/Daniel Cox (c), Mike & Elvan Habicht (br); **87:** OSF/Richard Alan Wood; **88:** OSF/ Frank Huber (bc), Tom Ulrich; **89:** OSF/E. R Degginer (b), OKAPIA (cra), Ronald Toms (c); **90:** OSF/Lon E. Lauber (c), Richard Kolar (cla), Stan Osolinski (tl); **90–91:** Tom Ulrich (b); **91:** OSF/Alan & Sandy Carey (tl), Leonard Lee Rue (cla), Mark Hamblin (tr), Richard Packwood (cbr), Rick Price (crb), Ted Levin (cfr); **91:** Andrey Zvoznikov (br); **92:** OSF/Bennett Productions/ SAL (bl), Daniel and Julie Cox (cal); **92–93:** Norbert Rosing; **93:** OSF/Dan Guravich (tc), Daniel Cox (c), Sue Flood (bc); **95–95:** OSF/ L.L Rhodes; **96:** OSF/Chris Catton (crb), Konrad Wothe (cfr); **97:** OSF/Michael Fogden (tc), Rick Price/SAL (crb), Wendy Shattil & Bob Rozinski (cra); **98:** OSF/Berndt Fischer (tl), Colin Monteath (cla); **98–99:** Wendy Shattil & Bob Rozinski; **99:** OSF/M Wendler/OKAPIA (tr), Michael Dick/AA (cla); **100–101:** Corbis/Tom Brakefield (l); **101:** Corbis/Martin Harvey; Gallo Images (tr); **101:** OSF (br), Alan Root/SAL (cfl); **102:** OSF/Chris Sharp (tl), Des & Jen Bartlett/ SAL (bl), Judd Cooney (c), Richard Packwood (cla), (cl), Stan Osolinski (cbl); **102–103:** OSF/ Rick Price/SAL; **103:** OSF/Partridge Films Ltd (tr); **104:** N.H.P.A/Kevin Schafer (clb); **104:** OSF/Daniel Cox (cfl), Konrad Wothe (c); **105:** OSF/Alan Root/SAL (cfr), Colin Monteath (cra), J L Klein & M. L. Hubert/OKAPIA (c), Konrad Wothe (tr); Colin Monteath (cla); Michael Fogden (tl); **106–107:** OSF/Chris Catton; **107:** OSF/David Macdonald (br), Partridge Production Ltd. (tc); **108:** OSF/Mary Plage (tl), Nick Gordan (c),(bl); Partridge Films Ltd. (cla); **109:** Ardea London Ltd/K & L Laidler (tr); **109:** N.H.P.A/Stephen Dalton (br); **109:** OSF/Hans Reinhard/OKAPIA (ca), M. Wendler/ Okapia (bc), Partridge Films Ltd. (crb), Konrad Wothe (bl); **110–111:** OSF/Daniel Cox; **111:** OSF/David Cayless (br); **112:** OSF/David Tipling (cfl), Edward Parker (tl), Mike Powles (cla); **112–113:** Michael Fogden; **114:** OSF/ Brian Kenney (bc), Martyn Colbeck (br), (bcr), Shane Moore (c); **115:** Bruce Coleman Ltd/ Gunter Ziesler (cfr); **115:** N.H.P.A/Haroldo Palo Jr (br); **115:** OSF/Brian Kenney (ca), Edward Parker (cb); **116:** Derek Harvey (tcr); **116:** OSF/ Manfred Pfefferle (cr), Mickey Gibson/AA (b); **117:** OSF/Edward Parker; **118:** OSF/Norbert Wu (tl); **118–119:** OSF/Michael Fogden; **119:** OSF/Michael Fogden (br); **120:** OSF/Carol Farneti Foster; **121:** OSF/Aldo Brando (b), Mike Powles (tr); **122–123:** OSF/Carol Farneti Foster; **124:** Corbis/Tom Brakefield (b), W. Perry Conway (cl); **124:** OSF/Mike Powles (tc); **125:** OSF/Alan & Sandy Carey (tr), Jany Sauvanet/

OKAPIA (cr), (cb); **126–127:** Corbis; **128:** OSF/ Roger Brown (cb); **129:** Nature Picture Library Ltd/Peter Scoones (bl); **129:** OSF/Bert & Babs Wells (cra), David Kurl (cr), Eric Woods (br); **130:** Corbis/Charles Philip (br); **130:** OSF/Bert & Babs Wells (ca), Kathie Atkinson (tl), (c), Michael Fogden (cfl), Roger Brown (cla); **131:** OSF/Daniel Cox (cra), David Kurl (b), Richard Packwood (tl); **132:** OSF/Grant Dixon/HH (tl), Kathie Atkinson (bl), Nigel Westwood/SAL (tr); **133:** Auscape/D. Parer & E. Parer-Cook (tr); **133:** OSF/Eric Woods (b), Konrad Wothe (cal); **134:** OSF/Des & Jen Bartlett/SAL (ca), Kathie Atkinson (cla), (cfl), Roger Brown (tl); **134–135:** Roger Brown (b); **135:** John Cancalosi (br); **135:** Dr C Andrew Henley-Larus (bcr), Dr C Andrew Henley-Larus (bc); **135:** OSF/David Curl (tr), Kathie Atkinson (tcl); **136:** Ardea London Ltd/Hans & Judy Beste (clb); **136:** Corbis/Joe McDonald (br), Kevin Schafer (crb); **136:** OSF/ Matthias Breiter (cla); Roger Brown (tl); **137:** Nature Picture Library Ltd/Peter Scoones; **138:** OSF/Bert & Babs Wells (cra), Chris Perrins (cla), Des & Jen Bartlett/SAL (bl), Kathie Atkinson (cfl), Lloyd Nielsen (tl); **139:** Peter Cross (br); **139:** Foto Natura/Martin Harvey (cfr); **139:** OSF/Bert & Babs Wells (t); **140:** OSF/ Hans & Judy Beste/AA (tc), Hans & Judy Bester/AA (c), Robin Bush (br); **141:** OSF/Daniel Cox; **142–143:** OSF/Bert & Babs Wells; **144:** OSF/ Stanley Breeden; **145:** Chris Mattison Nature Photographics/Martin Withers (tr); **145:** N.H.P.A/Ralph & Daphne Keller (cra); **145:** OSF/Belinda Wright (bl), Bert & Babs Wells (tc), Konrad Wothe (cr); **145:** Dave Watts (crb), (br); **146–147:** OSF/Clive Bromhall; **148:** OSF/Martyn Colbeck (ca); **148:** Getty Images/Art Wolfe (br); **149:** Corbis/Nevada Wier (tr); **149:** OSF/Ajay Desai (cr), Richard Packwood (bl); **150:** OSF/Colin Monteath (tl), David Tipling (car), (cfl); George Reszeter (cla); **150–151:** Getty Images/Art Wolfe (b); **151:** OSF/Dinodia Picture Agency (tr), E R Degginer/ AA (cla); **152:** Corbis/Nevada Wier; **153:** OSF/David Tipling (ca), Konrad Wothe (cfr), Mike Powles (cra), Richard Davies (tr), Stan Osolinski (br); **153:** Otto Pfister (bcl), (bcr); **153:** Dave Watts (bc); **154–155:** OSF/Martyn Colbeck; **156:** Ardea London Ltd/Kenneth W. Fink (cra); **156:** OSF/Henry M. Mix (tl), (cla); **156–157:** Martyn Colbeck, (b); **156:** Richard Matthews (clb); **157:** OSF/Colin Monteath (tr); **158:** Peter Cross (crb); **158:** OSF/Daniel Cox (cla), Manfred Pfefferle (tl), Stan Osolinski (cfl); **158–159:** Richard Packwood (b); **159:** BIOS Photo/GV-PRESS/Seitre (br); **159:** OSF/David Cayless (tr); Stan Osolinksi (cfr); **159:** Still Pictures/Rowland Seitre (crb); **160:** FLPA - Images of nature/Terry Whittaker (bl.); **160–161:** OSF/Jon Downer (t); **161:** OSF/Doug Allan (cfr), Konrad Wothe (br); **162:** OSF/Mella Panzella/AA; **163:** BIOS Photo/Rowland Seitre (c), **163:** OSF/Michael Dick/AA (tr), Stan Osolinski (bl), (bcr); **164–165:** OSF/Ajay Desai (b), Colin Monteath (tl), Dinodia Picture Agency (cla), (cfl); **165:** Ardea London Ltd/Pat Morris (cfr); **165:** Bruce Coleman Ltd/Alain Compost (tr), (crb); **165:** F. Jack Jackson (br); **165:** Mark

Kostich Photography (cra); **165:** OSF/Krupaker Senani (cla); **166:** OSF/Peter Weimann (tr), Stan Osolinski (ca); **167:** OSF/Keith & Liz Laidler (c), Mike Powles (clb); **168:** OSF/Bob Bennett (tc), **168–169:** OSF/Mike Hill (c); **169:** OSF/Barbara Von Hoffman (crb), Daniel Cox (cra); **170:** OSF/ Alan Nelson (cfl), Jorge Sierra Antinolo (bc), Kathie Atkinson (tr), Stan Osolinksi (tl); **171:** OSF/David Haring; **172:** OSF/Partridge Prod Ltd (tc), Partridge Prod. Ltd (b); **173:** OSF/Hans & Judy Beste/AA (r), Vivek Sinha/SAL (l); **174–175:** OSF/Frithjof Skibbe; **176:** Corbis/Pat Jerrold; Papilio (br); **176:** OSF/Chris Knight/SAL (ca); **177:** OSF/Berndt Fischer (cr), Gordon Maclean (tr); **178:** OSF/Berndt Fischer (cla), Chris Knight/SAL (b), Simon Tupper (tl); **179:** OSF (br), Bill Paton/SAL (tr), Carlos Sanchez Alonso (cla), Mark Hamblin (cl), Tony Tilford (cfr); **180:** OSF/Daniel Cox (cbr), Daniel Valla/ SAL (tl), Harold Taylor (cla), Konrad Wothe (cfl), Mark Hamblin (bl), Mike Brown (car); **180:** Dave Watts (bcr); **181:** Corbis/Pat Jerrold; Papilio (br) **181:** OSF/David Tipling (tl); **182:** OSF/Gordon Maclean; **183:** OSF (cl), Berndt Fischer (cra), Doug Allen (tr), Niall Benvie (cfr); **184:** OSF/ Mike Birkhead (bcl), Norbert Rosing (car), Paolo Fioratti (cfl), Philippe Henry (cla), Richard Packwood (tl), Tony Bomford/SAL (cbl.); **185:** OSF/Berndt Fischer (c), Peter Weimann/AA (br); **186–187:** OSF/Keith Ringland; **188:** OSF (bcl), Mark Hamblin (c); **189:** OSF/Ian West (cal), Owen Newman (bl), Philippe Henry (tr); **190–191:** Getty/Digital Vision; **192**: OSF/Jeff Foott/Okapia (c); **193:** OSF/Ben Osborne (tcl), Tui De Roy (bl), Dinodia Picture Agency (cla), Robin Bush (tl); **194–195:** OSF/Claude Steelman/SAL; Jeff Foott/Okapia; **195:** William Bernard Photography/William Bernard (tr); **195:** OSF/Gerard Soury (cr), Jeff Foott/Okapia (crb), Richard Herrmann (tcr); **196:** Peter Cross (tr); **197:** OSF/Jeff Foott/OKAPIA (cfr), Lon E Lauber (tr); **198:** OSF/Gerard Soury (clb), (car), Peter Ryley (tl); **198–199:** Tui De Roy (bl); **199:** OSF/Daniel Cox (br), Gerard Soury (tc), (cla); **200–201:** OSF/Duncan Murrell; **202:** Ardea London Ltd/Andrea Florence (bcl); **202–203:** OSF/David Fleetham, Gerard Soury (cl), (bc), Tony Bomford (tr); **203:** OSF/ Gerard Soury (cr), Howard Hall (br), Richard Kolar/AA (tr); **204:** OSF/Ben Osborne (cla), (cfl), Daniel Cox (tl), Doug Allan (ca), (bl), (br); **205:** OSF/Ben Osborne, Rick Price/SAL (br); **206–207:** OSF/ Stan Osolinski, (b); **207:** OSF/Lon E Lauber (tc), Norbert Rosing (cfr).

All other images © Dorling Kindersley. For further information see: www.dkimages.com

Every effort has been made to trace the copyright holders and we apologise in advance for any unintentional omissions. We would be pleased to insert the appropriate acknowledgement in any subsequent edition of this publication.